The Pro/ENGINEER® Exercise Book

The OnWord Press Development Team

The Pro/ENGINEER® Exercise Book

By the OnWord Press Development Team

Published by
OnWord Press
2530 Camino Entrada
Santa Fe, NM 87505-4835 USA

Copyright © 1994 Texas Instruments, Incorporated
SAN 694-0269
First Edition, 1994

10 9 8 7 6 5 4 3 2

Printed in the United States of America

Library of Congress Cataloging-in-Publication Data

OnWord Press Development Team
The Pro/ENGINEER Exercise Book

Includes index.

1. Pro/ENGINEER (computer software) 2. Mechanical computer-aided design I. Title
93-85032

ISBN 1-56690-037-9

Trademarks

Pro/ENGINEER®, Pro/TABLE®, and Pro/MANUFACTURE® are registered trademarks of Parametric Technology Corporation. OnWord Press is a registered trademark of High Mountain Press, Inc. Other products and services are mentioned in this book that are either trademarks or registered trademarks of their respective companies. OnWord Press and the development team make no claim to these marks.

Warning and Disclaimer

This book is designed to provide information about Pro/ENGINEER. Every effort has been made to make this book complete and as accurate as possible; however, no warranty or fitness is implied.
The information is provided on an "as-is" basis. OnWord Press shall have neither liability nor responsibility to any person or entity with respect to any loss or damages in connection with or rising from the information contained in this book.

About the Author

Bill Paul, P.E., is Manager of CAD/CAE training at the Defense Systems and Electronics Group of Texas Instruments, Incorporated. He has a BS and MS from Mississippi State University. Bill has over 30 years experience in mechanical design, analysis, training, and CAD.

Acknowledgments

We are grateful for the assistance of many who have helped in the preparation of this volume. The HRD staff at the Texas Instruments Learning Institute was instrumental in the development of the Pro/ENGINEER training material on which this book is based. In particular, our thanks go to Steve Brown and Dwight Saffel for their invaluable help. We also wish to thank Dave Rakestraw for the preparation of the answer key for the exercises.

Book Production

This book was produced in Microsoft Word for Windows 2.0b. The cover design is by Lynne Egensteiner, using QuarkXpress 3.11 and Aldus FreeHand 3.0.

OnWord Press

OnWord Press is dedicated to the fine art of professional documentation. In addition to the development team who edited the material for this book, other members of the OnWord Press team contributed to making this book. Thanks to the following people and the other members of the OnWord Press team who contributed to the production and distribution of this book.

> Dan Raker, Publisher
> Kate Hayward, Director of Operations
> Gary Lange, Manager of Contracts and Administration
> David Talbott, Acquisitions Editor
> Carol Lebya, Production Manager
> Frank Conforti, Managing Editor
> Margaret Burns, Project and Production Editor
> Lynne Egensteiner, Designer
> Tierney Tully and Bob Leyba, Production Assistants

Conventions Used in this Book

The following conventions were used throughout this book:
Menu names appear in all capital letters, e.g., MODIFY SCHEME.
Commands and submenu selections appear in italic type, e.g., *Reroute* command.

Companion Disk Installation

The bonus companion disk is attached to the inside back cover of this book. The disk contain exercise drawings for you to use as you go through the exercises. The disk is a high-density, 1.44 Mb DOS format disk.

Installation on a Windows NT Workstation

To install the disk,

1. Insert the disk into your floppy drive.
2. Using File Manager, create a directory on your hard drive.
3. Copy all of the files from the disk to the new directory.

Installation on a UNIX Workstation that Can Read a DOS Disk

1. Insert the disk into your floppy drive.
2. Mount the floppy drive if it is not already mounted.
3. Create a directory in your home directory.

4. Using a "dos2unix" transfer method, copy the files from the disk to the new directory.

Consult your manual or system administrator if you are unsure of how to perform any of these operations.

Installation on a UNIX Workstation that Cannot Read a DOS Disk

If you have a workstation that cannot read DOS disks, but is connected to a network with a PC, you can transfer the files.

1. Insert the disk into your floppy drive on the PC.
2. Copy the files from the disk to a temporary directory on the PC.
3. Create a directory in your home directory on the workstation.
4. Transfer the files to the workstation via TCP/IP connection, network mail, or any other available method.
5. Convert the files from DOS format files to UNIX format files if the conversion did not already take place during the transfer process.

Consult your manual or system administrator if you are unsure of how to perform any of these operations.

The Pro/ENGINEER® Exercise Book

The OnWord Press Development Team

Table of Contents

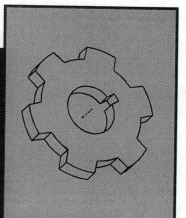

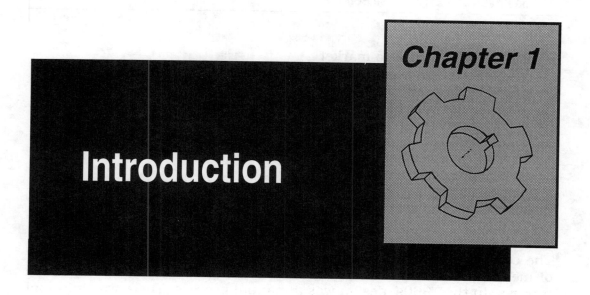

Chapter 1

Introduction

Concurrent engineering is being implemented by many companies in order to compete more efficiently in world markets. Pro/ENGINEER is an excellent tool to implement concurrent engineering into the mechanical design process. It is a parametric-based solid modeling system. This means the models are dimension driven, easy to change, and have "intelligence."

In this chapter we cover

- ❑ Concurrent engineering
- ❑ Mechanical design process automation objectives
- ❑ Pro/ENGINEER and mechanical process automation requirements

Concurrent Engineering

The following definition is taken from the Institute for Defense Analysis:

> Concurrent engineering is a systematic approach to the integrated, concurrent design of products and their related processes, including manufacture and support.

In summary, *concurrent engineering* simply means *communication* between all organizations during product development.

The term *concurrent engineering* is thrown around quite freely as a buzzword. What does it really mean in terms of benefits? There are several positive results that can be realized through concurrent engineering. They are as follows:

1. Shorter time to market
2. Lower product development costs
3. Higher product quality
4. Lower manufacturing costs
5. Lower testing costs
6. Reduced service costs
7. Enhanced competitiveness
8. Improved profit margins

There really is no choice; we need to improve in these areas or we will lose out to the competitors who choose to improve their design process.

The concept of concurrent engineering is not new. A look at the history of mechanical concurrent engineering will show us the evolution of the process. In the 1960s, design was a manual process. There was informal consulting between the designer, drafters, and manufacturers. At the time it was not known as concurrent engineering, although the results were similar.

During the 1980s, wire frame CAD was implemented. This created a semiautomated drafting-based design environment. Instead of concurrent design, a formal review procedure was installed between designer and drafter. We are currently in this phase of design automation. This process has created the use of multiple databases in design, analysis, and manufacturing (see Figure 1-1). The use of multiple databases has created many problems during the life of the product. The problems can include the following:

1. The drawing is the master design, resulting in 50% to 90% of all significant drawings having missing or incorrect dimensions.

2. Multi-database product definition environment is the basis for reduced product quality, resulting in a high number of engineering changes.

3. Concurrent engineering is an after-the-fact design review process, resulting in four or more drawing iterations per part.

4. Analysis and manufacturing automation is neutralized by the large amount of data preparation required.

Mechanical Design Process Automation Objectives

To improve these statistics a mechanical design process automation objective can be established. The objective is to create an engineering-based design environment, using a single product definition database for

MECHANICAL DESIGN PROCESS
UNDERLYING PROBLEM: MULTIPLE DESIGN REPRESENTATIONS

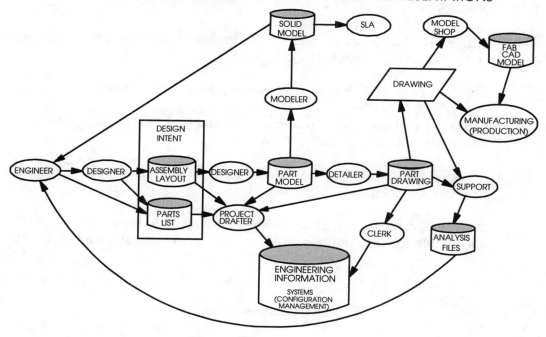

Figure 1-1.

all design, drafting, analysis, and manufacturing. This allows for concurrent data flow along with the following improvements:

1. The assembly model database is the master design, resulting in approximately 0% of engineering released drawings having missing or incorrect dimensions.

2. Single product definition database is the basis for all analysis and manufacturing functions, resulting in 50% to 70% reduction in engineering changes.

3. Concurrent engineering is a team design environment, not a review process, resulting in one drawing cycle per part.

4. Analysis and manufacturing use the design database directly for all process planning operations.

Pro/ENGINEER and Mechanical Design Process Automation Requirements

Pro/ENGINEER fulfills the following critical requirements for an integrated product development outline:

1. Efficient computer hardware/communications network.

2. A design process that creates a product database using the concurrent engineering philosophy.

3. Acceptable CAE/CAD/CAM functionality for each of the critical functional tasks.

4. A design database that is complete, unambiguous, and directly accessible for all engineering, analysis, and manufacturing operations.

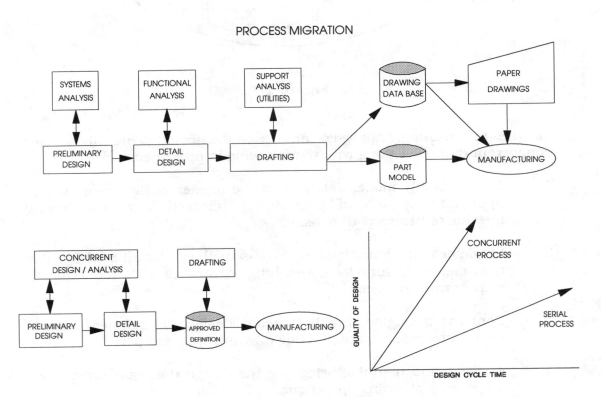

Figure 1-2.

CONCURRENT ENGINEERING

WORKSTATION/INTEGRATION REQUIREMENTS

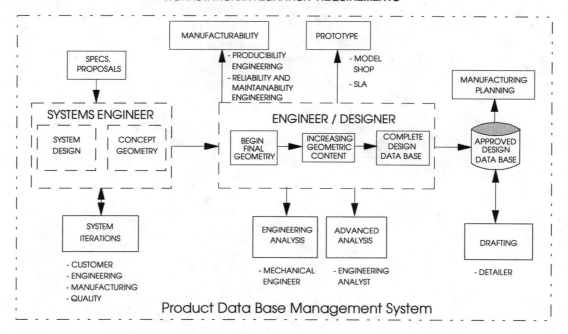

Figure 1-3. Product database management system.

5. A fully associative design/drawing database.

6. A full range of engineering analysis capability, including geometric (weight, center of gravity), finite element model, and tolerance.

7. A product data management system to manage the flow of data within for the concurrent process.

In the following chapters, you will learn how to use Pro/ENGINEER to accomplish the concurrent engineering process.

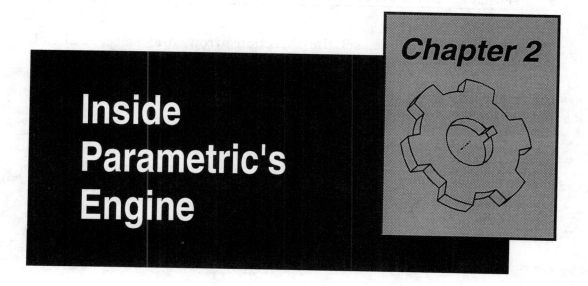

Inside Parametric's Engine

Chapter 2

To understand how Pro/ENGINEER constructs a part, you must have a good understanding of how Pro/ENGINEER creates each feature. You also need to be aware of techniques used in creating these features.

In this chapter we cover

- ❏ What a feature is
- ❏ How feature information and parts are stored
- ❏ The Sketcher environment
- ❏ Rules and assumptions for regenerating section Sketches
- ❏ Unsuccessful regeneration of a Sketch
- ❏ Setting Sketcher accuracy
- ❏ Special geometry in Sketcher

What Is a Feature?

Pro/ENGINEER is a *feature*-based solid modeler. This means that every part is made up of a series of features. Each feature changes the appearance of and adds information to the model. When a feature is created, all of the information that defines the feature is recorded and filed in memory. It can be changed or modified as the design progresses.

Take, for instance, a "D-shaped" cut created in a simple block as in Figure 2-1. To create this cut, select the following menu items: *Feature, Create, Cut, Extrude/Solid/Done, Single/Done,* and *Thru-All/Done.* Then select a Sketching plane, set the Direction arrow, and select a sketch Reference plane (Top, Bottom, Left, or Right). Only after this is it possible to sketch the cut and dimension it. In this process much information is given to Pro/ENGINEER. This information remains with the cut forever unless the cut is deleted.

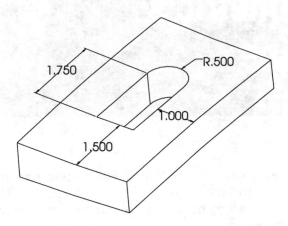

Figure 2-1. A D-shaped cut.

In review, we just created an extruded cut, *Thru all* the part traveling in one direction from the sketching plane.

In Pro/ENGINEER a feature is any Hole, Shaft, Round, Chamfer, Slot, Cut, Protrusion, Neck, Flange, Rib, Shell, or Pipe. Datums (Planes, Axes, Curves, Points, Coordinate Systems, Graphs) are features, but are used as references in geometry construction and placement.

How Is Feature Information Stored?

Let's take a look at how Pro/ENGINEER stores information on each feature. The *Feature Form* was developed to help us understand this process better. This form contains all of the important information about a feature. If we could look inside a Pro/ENGINEER part file, we might see the "D-Shaped" cut stored in the format shown in Figure 2-2.

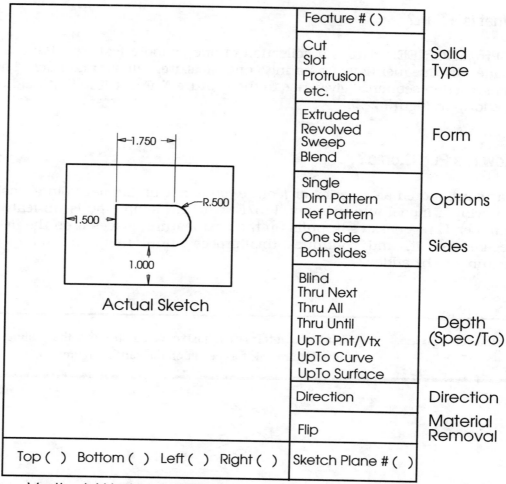

Figure 2-2. The feature form.

On this form each space that contains information is called a *field*. On the right side of the form are fields that contain feature construction information such as *Cut, Extruded, Thru-All, Direction,* and so forth. Across the bottom of the form is the feature reference information such as the surfaces chosen for both the Sketching Plane and the Reference Plane. In the center of the form is the sketch itself. The sketch has the largest field on the form and contains the sketch and all associated dimensions and alignments. It is important to remember that a feature is not just the sketch but is all of the information in all of the fields, including the sketch itself.

What Is a Part?

A Pro/ENGINEER part is a collection of one or more features. Parts are made by sequentially creating one feature after another. This construction sequence gives rise to the Feature Number (the first field on the form in Figure 2-2).

How Is a Part Stored?

Parts are stored as a series of *feature forms* not unlike pages in a book. To bring a model into memory, Pro/ENGINEER simply pulls up feature number 1, regenerates it, pulls up the next feature, places it on the first, regenerates it, and continues this process until there are no more features to be added.

Exercise 2-1:   Using the information just covered, fill out the feature form, Figure 2-4, for the cut on the part in Figure 2-3.

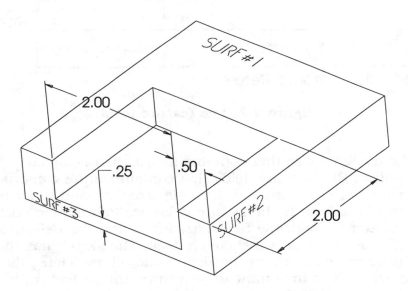

Figure 2-3.

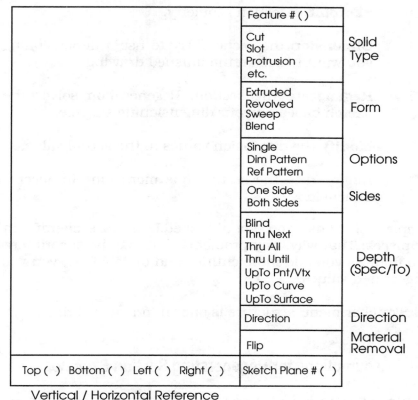

Feature # ( )	
Cut Slot Protrusion etc.	Solid Type
Extruded Revolved Sweep Blend	Form
Single Dim Pattern Ref Pattern	Options
One Side Both Sides	Sides
Blind Thru Next Thru All Thru Until UpTo Pnt/Vtx UpTo Curve UpTo Surface	Depth (Spec/To)
Direction	Direction
Flip	Material Removal
Sketch Plane # ( )	

Top ( ) Bottom ( ) Left ( ) Right ( )

Vertical / Horizontal Reference

Figure 2-4.

The Sketcher Environment

As you may know by now, Sketcher techniques are used in many areas of Pro/ENGINEER. The premise of Sketcher, and of Pro/ENGINEER, is to allow the quick and simple creation of geometry for your model, i.e., most features are created from two-dimensional sketches. This means that a great deal of a designer's time is spent in the sketch mode. Sketcher requires you to create and dimension the geometry, but during the sketching process you don't have to be concerned with exact dimension values. You are probably going to modify the dimensions, and therein lies the reason for using parametric models.

Sketcher is an extremely powerful but complex tool, and you need to have a good working knowledge to make best use of it. In this section we discuss techniques, concepts, and tools for using the Sketcher. There are five basic steps to remember when you are using the Sketcher:

❏ **Sketch** the section geometry.

❏ **Dimension** the section. Try to use a dimensioning scheme you want to see on the finished drawing.

❏ **Regenerate** the section. Regeneration solves the section sketch based on your dimensioning scheme.

❏ **Modify** the dimension values to the actual values.

❏ **Regenerate** the section one more time to accept the new dimensions.

On complex sections, you should sketch and regenerate in several smaller pieces. That way, any problems that may be encountered will be easier to find. As you add more entities and dimensions, you will want to use exaggerated values.

The full Sketcher menu structure is shown in Figure 2-5.

Rules and Assumptions for Regenerating Section Sketches

During the regeneration of a section, the system checks to make sure that it understands your dimensioning scheme and that you have created a complete and independent set of parameters. It analyzes your section based on the geometry that you have sketched and the dimensions you have assigned. In the absence of explicit dimensions, implicit information or assumptions based on the sketch may be used. Table 2-1 presents implicit information or assumptions that Pro/ENGINEER uses to solve and regenerate a section.

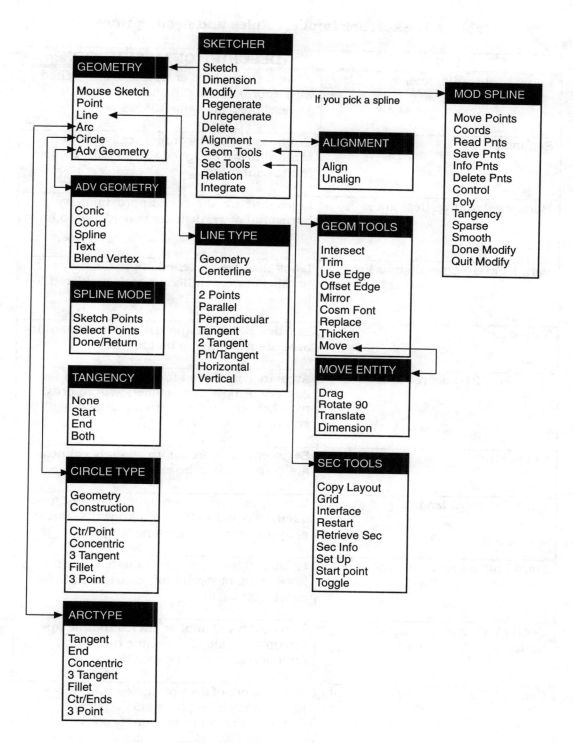

Figure 2-5. The Sketcher menu structure.

Table 2-1. Sketcher Implicit Rules and Assumptions

RULE	DESCRIPTION/ASSUMPTION
Equal radius/diameter	If two or more arcs or circles are sketched with approximately the same radius, they are assigned the same radius value
Symmetry	Entities sketched symmetrically about a centerline are assigned values with respect to the centerline
Horizontal and vertical lines	Lines that are sketched approximately horizontal or vertical are considered to be exactly so
Parallel and perpendicular lines	Lines that are sketched approximately parallel or perpendicular are considered to be exactly so
Tangency	Entities sketched approximately tangent to arcs are assumed to be tangent
90-, 180-, 270-degree arcs	Arcs are considered to be multiples of 90 degrees if they are sketched with approximately horizontal or vertical tangents at the end points
Colinearity	Segments that are approximately colinear are considered to be exactly so
Equal segment length	Segments of unknown length are assigned a length equal to that of a known segment of approximately the same length
Point entities lying on other entities	Point entities that approximately lie on lines, arcs, or circles are considered to be exactly on them
Centers lying on the same horizontal	Two centers of arcs or circles that lie approximately along the same horizontal direction are set to be exactly so
Center lying on the same vertical	Two centers of arcs or circles that lie approximately along the same vertical direction are set to be exactly so

Exercise 2-2: Using the information in Table 2-1, sketch the following sections and try to determine the Sketcher assumptions made.

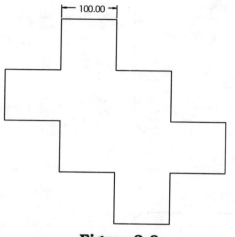

Figure 2-6.

1. _____

2. _____

3. _____

4. _____

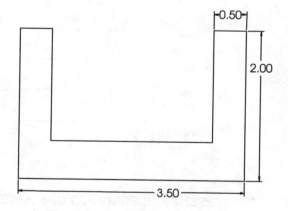

Figure 2-7.

1. _____

2. _____

3. _____

4. _____

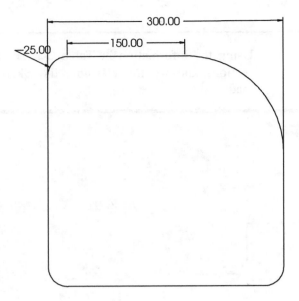

Figure 2-8.

1. _____

2. _____

3. _____

4. _____

5. _____

Unsuccessful Regeneration of a Sketch

If a section cannot be solved using the dimensioning scheme and implicit sketched information, Pro/ENGINEER displays a message and highlights the error. There are five basic categories of errors:

- ❑ **Sketch does not communicate intent:** If you do not sketch within reason, Pro/ENGINEER may fail to recognize your intent. Recover by sketching more clearly.

- ❑ **Underdimensioning:** If there is not enough information to solve the section, Pro/ENGINEER prompts with "Under-dimensioned Section" and highlights all unsolved vertices. Add the proper dimensions.

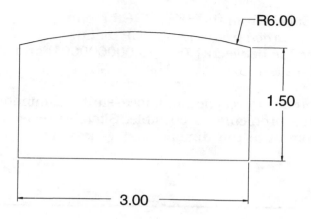

Figure 2-9.

☐ **Overdimensioning:** If there is too much dimensional information included in a section, Pro/ENGINEER solves the section and warns "Regeneration successful, EXTRA DIMENSIONS FOUND." If overdimensioning is required, first create a valid dimensioning scheme so the section regenerates successfully, then add any additional dimensions that you want to drive the model.

☐ **Zero length segments:** If a segment is sketched very small, Pro/ENGINEER prompts that there is a "Zero Length Segment" in the section.

☐ **Inappropriate sections:** If Pro/ENGINEER knows what a section will be used for, it will also check to make certain the section is appropriate. For example, if you create a revolved feature, Pro/ENGINEER checks to make certain that you have a centerline and that the section is closed.

Other problems may arise once the sketch regenerates successfully if an improper dimension is inserted. Figure 2-9 illustrates an easily regenerated sketch. However, if the diameter of the arc is changed to a value less than 3 inches, the geometry will be invalid and the section will not regenerate.

Setting Sketcher Accuracy

Modifying the Sketcher accuracy can help solve certain section regeneration problems. To change Sketcher accuracy:

1. Choose *Set Up* from the SKETCHER menu.
2. Choose *Accuracy* from the SET UP menu.
3. Enter a value between 1.0e-9 (0.000000001) and 1.0.
    The default is 1.0.

In order to get sketches to regenerate successfully, eliminate as many of the previously listed problems as possible. Sketch precisely, dimension carefully, and make sure the dimensional values match the sketched geometry.

Exercise 2-3:	Accuracy problems can be overcome in many instances by sketching the section using exaggerated dimensions, regenerating, and gradually approaching the actual dimension. Try this on the following exercise. Create and regenerate the sketch shown in Figure 2-10. Create all necessary construction geometry.

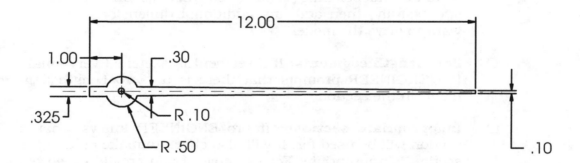

Figure 2-10.

Special Geometry in Sketcher

At this point, you should be well acquainted with creating geometry in Sketcher. There are, however, some geometric entities that can be used in various ways. The two most flexible entities are *Centerline* and *Point.*

Centerlines can be used to define the axis of revolution of a revolved feature, to define a line of symmetry within a section (mirror), or to create construction lines used for dimensioning or geometry alignment.

Points are added to a sketch to make it easier to dimension section geometry. Points can be sketched either on or off geometry; however, points placed off geometry are considered reference points. Points can also be selected for creating splines.

When points are added to a sketch, they must be dimensioned unless their location is implicitly defined. A point is implicitly defined when it is placed at the end of a line, spline, or arc, at the origin of arcs and circles, or at the intersection of two lines.

Exercise 2-4:   Using the information on points and centerlines above, and the information from Table 2-1, construct the sketch shown in Figure 2-11 and list the assumptions used.

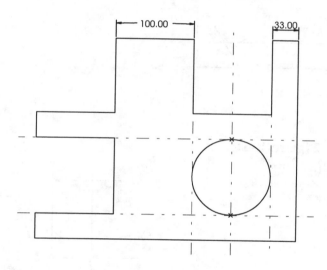

Figure 2-11.

1. _____

2. _____

3. _____

4. _____

5. _____

6. _____

Exercise 2-5:   Create and regenerate the sketches shown in Figures 2-12 and 2-13. Create all necessary construction geometry. Remember the assumptions and rules that were just covered (especially points).

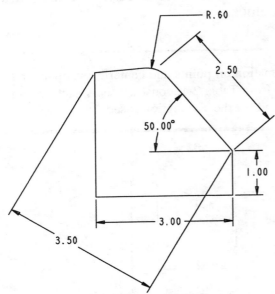

Figure 2-12.

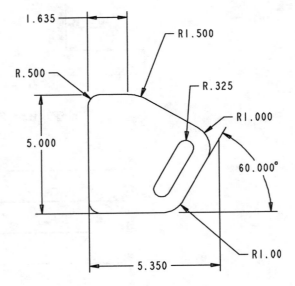

Figure 2-13.

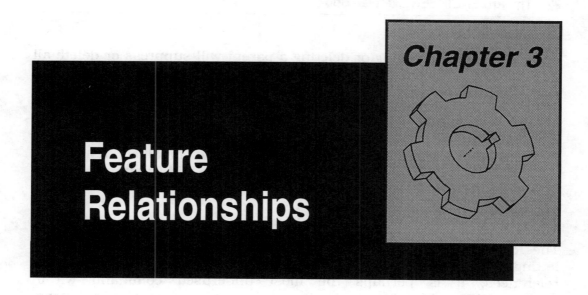

Chapter 3

Feature Relationships

Recognizing the relationships that exist between features and how these relationships can be modified or broken is critical for Pro/ENGINEER users who wish to create, update, and maintain parametric database integrity.

In this chapter we cover

- ❐ Parent/child relationships
- ❐ The *Reroute* command
- ❐ The *Redefine* command
- ❐ The *Reorder* command
- ❐ Constructing a Family Tree
- ❐ Constructing a local Family Tree

Parent/Child Relationships

Once a base feature is created, you can add other features to the model. The order of creating features is important. Any feature created after the base feature depends on earlier features in some way. Whenever a feature is created, it is considered the child of the feature(s) it is using for placement or dimensioning references. This is known as the parent/child relationship. Knowing what these relationships are is important when modifying a part.

There are three things to remember when working with parent/child relationships. These are

> ❑   Suppressing or deleting a parent will suppress or delete all of its children.

> ❑   Features cannot be reordered so that the child is created before *(older)* than the parent.

> ❑   Parent/child relationships can be modified using the *Reroute* or *Redefine* options from the FEAT menu.

The Reroute Command

The *Reroute* command is used to break parent/child relationships by letting you select new sketching, placement, and dimensioning references. It is perhaps the most underused command within Pro/ENGINEER. This may be because it has many uses. Unlike *Reorder* and *Redefine*, which have one specific use each, *Reroute* has many useful functions with which every Pro/ENGINEER user should be familiar.

Pro/ENGINEER keeps track of every surface, edge, or axis picked when creating a feature. How can these "references" be seen? For example, you may want to know what the sketching plane or the vertical or horizontal reference is for a specific feature. *Reroute* allows us to review, see, and even change these references.

How to Use Reroute

The *Reroute* command is always available in the FEAT menu. It also appears with the CHILD menu when a feature that has children is selected for deleting or suppressing.

Reroute from the FEAT Menu

To execute, choose *Reroute* from the FEAT menu; then pick the feature whose references are to be rerouted.

The system will ask, "*Do you want to roll back the part? [N].*" This will remove from the display any feature created after the feature being rerouted. It is a good idea to *always* "Roll Back the Part." This prevents a younger feature from being used as a reference, which would void the reroute.

When you enter "Y" <CR> to roll back the part, the system will display the original feature cross-section and all dimensions used to define the section. It will also highlight the sketching or placement plane, depending on the type of feature.

The *Reroute* function will highlight all feature references one by one (in the order they were selected). This includes all surfaces, edges, or axes chosen for creating or defining any of the following:

- ❏   Sketching plane
- ❏   Vertical or horizontal reference plane
- ❏   Dimensions
- ❏   Alignments
- ❏   Placement plane (for features like holes and shafts)
- ❏   Placement edge (for features like rounds and chamfers)
- ❏   Datums on the fly
- ❏   Datums

As each reference is highlighted, make a selection from the *Reroute* menu to select either an *Alternate* reference, the same reference *(Same Ref)*, request reference information *(Ref Info)*, or quit the reroute function *(Quit Reroute)*.

Note: Some references may not be able to be rerouted because suitable alternate references cannot be selected. After all possible references have been rerouted, the feature will regenerate. If regeneration is successful, the new parent/child relationship will be established. If not, the original references will be restored.

For those references that could not be rerouted, use the *Redefine* option.

Exercise 3-1:	In this exercise you will reroute the references for the slot feature located in the lower left of the part (*feature.prt* on the enclosed disk) shown in Figure 3-1.

To use the *Reroute* command successfully, you should have some knowledge of how the part is constructed. Use the *Regen Info* option from the INFO menu to determine the *part regeneration sequence*. The *Regen Info* option allows us to observe part regeneration from the base feature or from any feature specified. It also supplies the feature numbers. Continue stepping through the regeneration process until the information you need has been obtained.

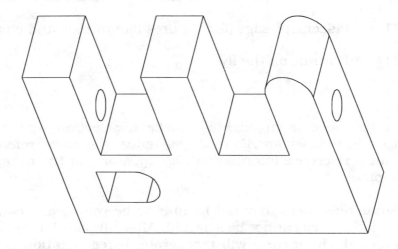

Figure 3-1.

Select *Feature* and *Reroute*. Select the feature to be rerouted and answer *yes* to roll back the part. Notice that the feature is highlighted along with the first reference, the features section sketch, and all dimensions. Select the desired options from the REROUTE menu for the highlighted reference. You will notice that some references cannot be rerouted. When this happens, we will complete the reroute and then use *Redefine* to complete the process.

The Redefine Command

Why redefine? The *Redefine* option allows you to modify how a feature is created. Which type of feature is selected will determine which definitions may be modified. For a through hole, the attributes may be changed. For a cut, you can redefine the section and the material being removed. *Redefine* will not change the sketch plane or the reference plane. In the following paragraphs we review some feature types and what options can be redefined.

How to Use Redefine

Choose *Redefine* from the FEAT menu and pick the feature to be redefined. Any feature can be redefined; however, the application changes with the type of feature. Following is a summary of how *Redefine* can be used on most of the common features.

Redefining Sketched Features

Cuts, slots, protrusions, sketched cosmetics, and sketched datum curves are all sketched features. Redefining allows you to change some or all of the following options:

Attributes	Retrieves the available depth and intersection attributes for redefining.
Direction	Allows the direction in which the feature is created to be redefined.
Section	Retrieves the section sketches of the feature for redefining.
Flip	Allows the side to which material is added or removed to be changed.
References	Allows respecification of the feature's placement references (such as Until, From, and To surfaces).
Scheme	Allows the changing of the dimensioning scheme without revising the feature section. Using this, rather than Section, there is no chance that you could delete entities referenced by other features.

Scheme is not a subset of the *Section* option. When redefining with *Scheme*, you are only allowed to delete and add dimensions to the section while no other modifications to the section are possible.

Redefining Datum Planes

A datum's orientation (red/yellow sides) can be changed by redefining the *Direction*. Redefining the datum plane's *References* allows new options and references to be selected. Datum planes that are sketching planes created on the fly may only be redefined when they fail to regenerate.

A datum's attributes can be changed using the *Redefine* command. Datum attributes are the size constraints placed on datums when they are created. These include the following:

Default	The datum plane is sized to the model (part or assembly).
Fit Part	(Available in Assembly only) Sizes the datum plane to the selected part.
Fit Feature	Sizes the datum plane to a part or assembly feature.
Fit Surface	Sizes the datum plane to any surface.
Fit Edge	Sizes the datum plane to fit an edge.
Fit Axis	Sizes the datum plane to an axis.
Fit Radius	Sizes the datum plane to a specified radius, centering itself within the constraints of the model.

Redefining Features

When redefining a feature with a child that references the whole parent, the CHILD OPS menu appears with the following:

Delete All	Deletes all highlighted children.
Suspend All	Children features will not be deleted.
Child Info	Lists information about the children features.

Redefining Holes

The following can be redefined about a hole:

Scheme	Dimensions and dimensioning constraints.
Placement	MODIFY SCHEME menu. Change the location of the hole as defined by the new placement options *(Linear, Radial, Coaxial)*.
Section	Modify the dimensioning scheme of a blind hole cross-section.

The way in which a hole is dimensionally located on a part and the section dimensions of blind holes can be changed using *Redefine* and *Scheme*. A hole's depth is set by its *Attributes*, which are

- ❑ **Blind**
- ❑ **Thru-All**
- ❑ **Thru-Next**
- ❑ **Thru-Until**
- ❑ **From To**
- ❑ **One Side**
- ❑ **Both Sides**

Redefining Split and Regular Drafts

Regular Draft attributes and references can be redefined. When *Attributes* is selected for redefinition, the OPTIONS menu includes the following:

- ❑ **Constant/Variable**
- ❑ **Neutral/Ref & Neut**
- ❑ **Unmirrored/Mirrored**

When the *References* option is selected, an expanded FEATURE REFS menu appears with the following:

Add   Adds reference surfaces to the draft using the SURF/LOOP menu.

Remove   Removes reference surfaces from the draft using the SURF/LOOP menu.

Remove All   Removes all reference surfaces from the draft.

Replace   Replaces an existing draft reference surface.

ShowCurrRefs   Highlights all the feature reference surfaces of the current type.

The FEATURE REFS menu contains the following options for the addition and removal of draft points for Variable Drafts.

Add   Adds feature edge x plane reference points to a draft. New draft angle must be entered.

Remove   Removes an edge x plane reference point.

Remove All   Removes all edge x plane reference points.

Replace   Replaces an existing edge x plane reference point with another.

ShowCurrRefs   Highlights all edge x plane reference points used by the feature.

Sections, references, and schemes of split draft features can be redefined. The *Attributes* option is not available.

Patterns can be redefined if the pattern feature fails to regenerate because its parent reference becomes unusable. Select *Redefine* from the TRIM menu.

The attributes can be changed by choosing *Redefine*, then *Attributes*. The attribute used to define the hole originally will be preselected.

The direction from the placement plane that a hole penetrates the part can be changed by redefining the Direction.

Redefining Rounds

Edge Rounds

You can redefine the *Attributes* and *References* of edge rounds. The attributes specify whether the round is a constant or varying radius round. The references are the edges you selected for rounding. When redefining the references, Pro/ENGINEER will prompt you with a highlighted edge and the ADD RMV EDGE menu. For each edge, select *Add* or *Remove*. This lets you include new edges or remove existing edges from the round definition. When finished, choose *Done*.

Corner Rounds

You can change the values of the sphere radius and transition distances of intermediate surfaces of a corner round. The REDEFINE menu option *Corner Round* is accessible if the selected round feature contains corner rounds.

The Reorder Command

Through feature reorder, the order in which features are regenerated can be changed. Features can be moved forward or backward in the regeneration order list. Feature reorder cannot occur under the following conditions:

❑ Parents cannot be moved so that their regeneration occurs after that of their children.

❑ Children cannot be moved so that their regeneration occurs before that of their parents.

How to Use Reorder

Choose *Feature* from the PART menu, then choose *Reorder* from the FEAT menu. Pick whether the feature is to be *Earlier* or *Later* in the regeneration order, then pick the feature to be reordered. Pro/ENGINEER shows all possible feature numbers to which the feature can be reordered. Enter the desired feature number and press *Return*.

Reorder/Redefine/Reroute Summary Sheet

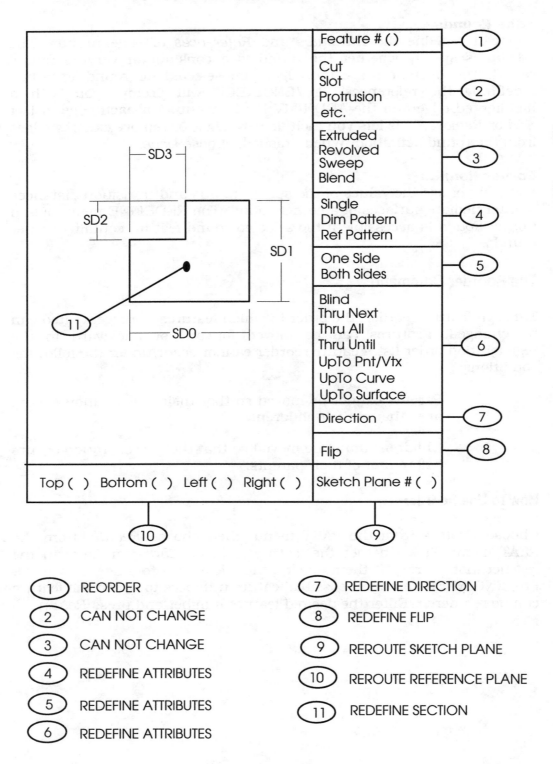

Figure 3-2.

Exercise 3-2: Using the *Reroute*, *Reorder*, or *Redefine* commands, make the part, *guideblk.prt* on the enclosed disk, look like the following drawing. *Do not delete any features.*

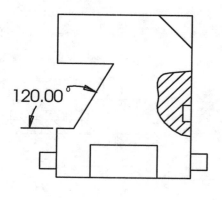
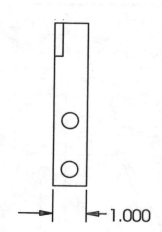
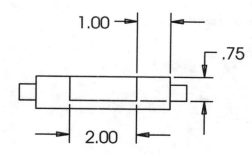

Figure 3-3.

Exercise 3-3: Make the changes to the part, *slavegr.prt* on the enclosed disk, to reflect the changes marked on the following drawing. Use *Regen Info* to get feature numbers. Use the feature numbers when selecting features to work on.

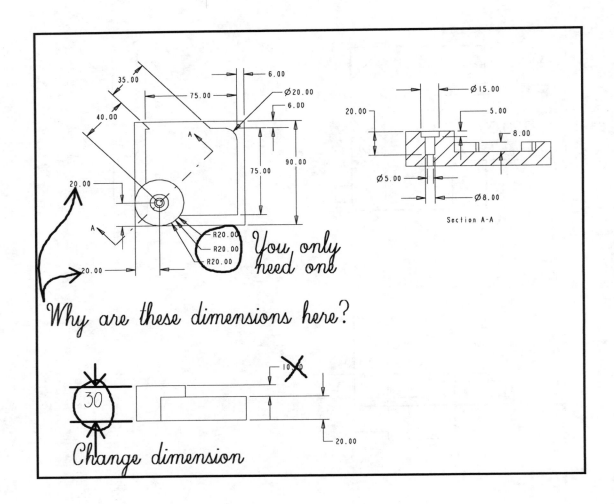

Figure 3-4.

Constructing a Family Tree

Parent/child relationships can be difficult to understand. The concepts in this section provide a way to deal with complicated parts and will allow you to clean up and correct parts that require complex rescheming and modification.

In order to understand the complex nature of rescheming a part, you must understand all of the ways in which parent/child relationships are created. Following is a list of the most common ways features can be children of other features.

❑ **Dimensions:** Anytime a feature is dimensioned from another feature, it becomes a child of the other feature.

❑ **Alignments:** While in Sketcher, if a sketched entity is aligned to a feature, it becomes a child of that feature.

❑ **Placement Plane:** When creating a hole or shaft, a placement plane is chosen for the location of the hole or shaft. This makes the hole or shaft a child of the feature whose surface was chosen.

❑ **Placement Edge:** In creating chamfers and rounds, edges are chosen. The chamfers and rounds are then children of the features whose edges were chosen.

❑ **Sketching Plane:** When creating a sketched feature, a sketch plane is chosen. This makes the sketched feature a child of the feature whose surface was selected as the sketch plane.

❑ **Vertical or Horizontal Reference:** When a sketched feature is created, not only is a sketching plane chosen, but a vertical or horizontal reference plane is chosen to position the part (Top, Bottom, Left, Right). The feature being created is a child of the feature whose surface was chosen as vertical or horizontal reference.

❑ **Constraint References:** Any time datum points, axes, or planes are created, other geometry is used to constrain the datum feature. Datum points, axes, and planes are, as a result, children of features used to construct them.

Now the question arises: how do you show these parent/child relationships in a clear and concise manner? The answer is by constructing a family tree showing the relationships between features, for all features in the part. In order for it to be useful, the family tree must contain two pieces of information: what features are children, and why they are children? The following example shows how such a parent/child relationship tree is developed for a simple part.

Example 1

Figure 3-5 shows a simple face-plate part (*faceplat.prt* on the enclosed disk) created in Pro/ENGINEER. The plate is made up of 11 features created in the following order:

1. Base Plate
2. Datum A
3. Datum B
4. Datum C
5. Hole 1
6. Hole 2
7. Hole 3
8. Hole 4
9. Mount Hole A
10. Mount Hole B
11. Cut Out

To understand how this part was created, the family tree is constructed, much like a flow chart, by following these steps:

1. On a piece of paper, draw a box with a number 1 to represent the base feature.

2. Reroute the next feature in the regeneration order (feature 2), using the *Reroute* option, to determine two things:

    a. What features were used to create this feature?
    b. What makes this feature a child of other features?

    Remember, when using the *Reroute* option, always roll back the part. Note the references as they highlight and what feature each is associated with.

3. On the line below the latest parent feature, draw a box with the feature number of the feature that was just rerouted.

4.   Draw lines connecting this box to all the parents indicated.

5.   Along each connecting line, note the conditions that make this feature a child of the other (refer to the message line).

6.   Repeat steps 2 through 5 for each feature in order.

7.   When all features have been added, stop.

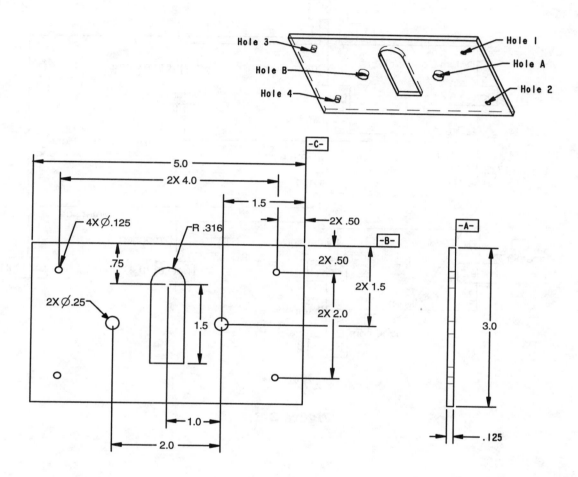

Figure 3-5.

Figure 3-6 shows what the Family Tree looks like for *faceplat.prt*, which can be found on the enclosed disk.

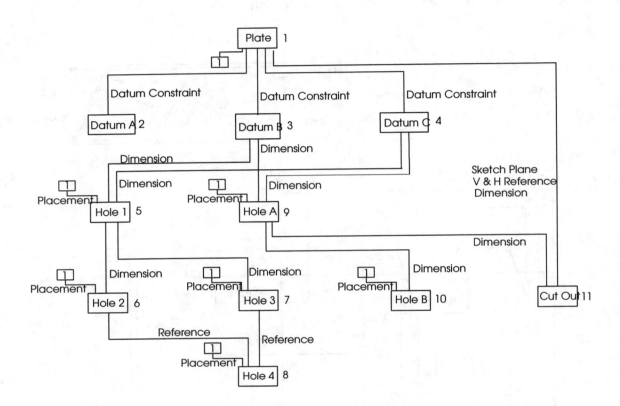

Figure 3-6.

Exercise 3-4:	1. What will happen if feature 2 is deleted?
	2. If we wish to delete feature 3 and not delete feature 11 at the same time, what must we do?
	3. What command (*Reroute*, *Reorder*, or *Redefine*) would you use to break the parent/child relationship between features 3 and 11?
	4. How far back can feature 9 be reordered?

1._____

2._____

3._____

4._____

Exercise 3-5: Using the space provided and following the steps described in Example 1, create a family tree for the part *vice_bas.prt*, shown in Figure 3-7.

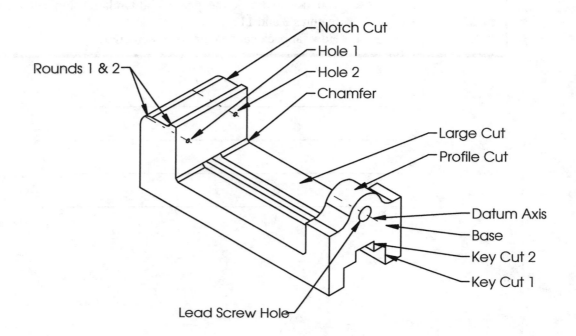

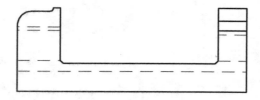
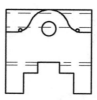

Figure 3-7.

Use this page to construct the family tree for the *vice_bas.prt*. Start with the base.

```
┌─────────────┐
│  Base (1)   │
└─────────────┘
```

Is this the best possible scheme for this part? Could you improve on it?

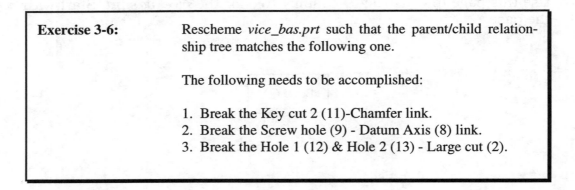

Exercise 3-6:   Rescheme *vice_bas.prt* such that the parent/child relationship tree matches the following one.

The following needs to be accomplished:

1. Break the Key cut 2 (11)-Chamfer link.
2. Break the Screw hole (9) - Datum Axis (8) link.
3. Break the Hole 1 (12) & Hole 2 (13) - Large cut (2).

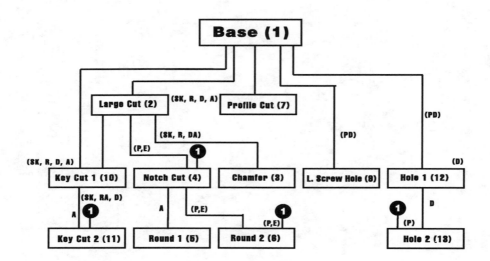

Figure 3-8.

Constructing a Local Family Tree

As we have just seen, the construction of the family tree diagram is useful when updating or modifying a part. It is much easier to rescheme or delete a feature once the feature's relationship to other features can be identified. However, it is not realistic to create a family tree diagram for every part. Family tree diagrams are cumbersome for parts with more than 15 or 20 features. Furthermore, each time a feature is changed or deleted, the diagram must be updated. For this reason, we need to explore methods for creating family tree diagrams that show only the area of concern, thus a *local family tree.*

Once we determine what changes are to be made to the model, we need to determine how the changes affect other features. One easy way to find all of a feature's children is simply to *Suppress* the feature. When suppressing a feature, all of the children will highlight blue one at a time, allowing us to identify each of them. To take full advantage of this, let's discuss what options are available while using the *Suppress* command.

When *Suppress* is chosen from the FEAT menu, the following options are available:

Normal	Suppresses the selected feature.
Clip	Suppresses the selected feature and all features created *sequentially* after this feature.
Unrelated	Suppresses all features other than those selected and their parents.
Quit Del/Sup	Quits out of the suppress process.

Once a feature is chosen to be suppressed by any of the three previously mentioned methods, other features will begin to highlight in blue. The statement is then presented: *Select option for the child in blue.* The following options are available:

Show Ref	Shows references for the child selected.
Reroute	Reroutes the references of the highlighted child.
Modify Scheme	Modifies the dimensioning scheme of the highlighted child.
Suppress	Suppresses the highlighted child.

Suppress All   Suppresses all the children.

Suspend   Keeps this child.

Suspend All   Keeps all children.

Info   Shows information about the highlighted child.

Quit   Quits out of/exit the process.

Exercise 3-7:   Gather information and build a local family tree for the center protrusion of *link_hol.prt,* found on the enclosed disk, shown in Figure 3-10.

1. First gather information about this feature. Select *Info* from the PART menu.
2. Select *Feat Info* from the INFO menu and select the center protrusion.
3. From the Information Window, extract the feature number and note the number of children. Exit the Information Window.
4. Select *Feature* from the PART menu and then select *Suppress* from the FEAT menu.
5. Select the feature to suppress. In this case, select the center protrusion.
6. As each child is highlighted in blue, select *Info* and note the feature number. Let's begin the family tree (see Figure 3-9).
7. Exit the Information Window and select *Suppress.* The next child highlights. Get the information on this child, add the information to the next box of the family tree, and suppress the child.
8. Continue the process until all children have been charted.
9. Quit the *Suppress* function.

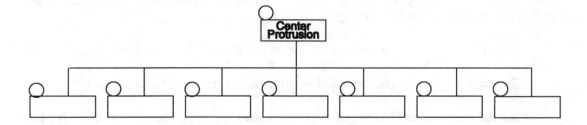

Figure 3-9.

Now the *Reroute* option can be used to reroute each feature that is related (a child) to the center protrusion. The protrusion can then be deleted or suppressed.

In many cases a family tree is not necessary at all. Even in cases when creating a local family tree is necessary, seldom does every parent and child feature need to be added. Experience is the best guide as to how many features must be added to the tree.

Exercise 3-8:	You have just been given the task to modify the part *link_hol.prt*, Figure 3-10, to incorporate two changes listed below. Use the local family tree that we just developed to assist you in this process. *Do not at any time delete any features other than the one listed.*

1. Delete the centermost of the three posts.
2. Make the square cut in the posts stop as it passes through the cut in the second post.

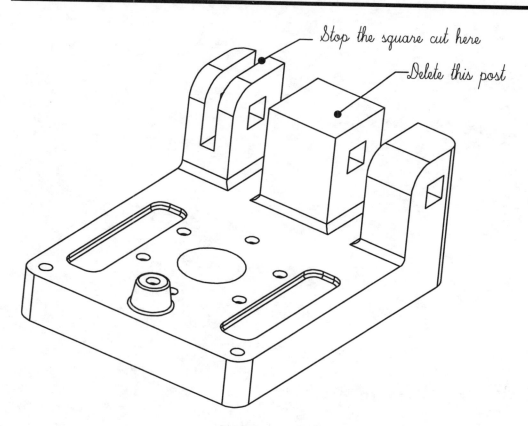

Stop the square cut here

Delete this post

Figure 3-10.

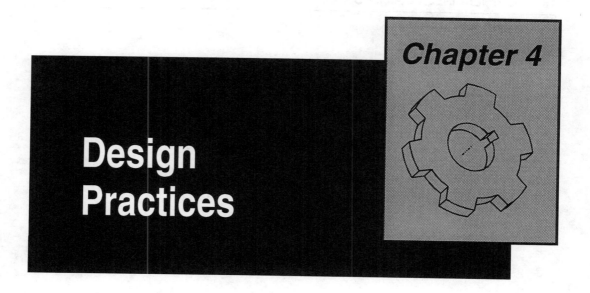

Design
Practices

ith a good working understanding of how a feature is stored in the part file, we can now concentrate on how to avoid trouble when creating features. In this chapter we cover the following:

❑   Datums on the fly

❑   Attributes for sketched features

❑   Fill-ins/Cut-off

❑   Reference planes

Datums on the Fly

What are datums on the fly? When are they available?

When creating any sketched feature, a Sketching plane and a Reference plane must be chosen. When appropriate, Pro/ENGINEER lets you pick existing datum planes or create datum planes using the *Make Datum* option. If *Make Datum* is chosen, the resulting datum is known as a datum on the fly. Choosing *Make Datum* tells Pro/ENGINEER that the datums are needed only for the construction of this feature. Datum planes that are created during feature creation become part of that feature. They become invisible after the feature has been created and any associated dimensions are included with those of the feature.

Figure 4-1 shows a protrusion sketched on a datum on the fly. The datum in Figure 4-1 is offset from DTM1 by a dimension of 5 inches. When the feature is finished the datum disappears. Figure 4-2 shows that the offset dimension to the datum has been "internalized" in the feature (the protrusion). In other words, when creating a feature, any dimension that locates a datum on the fly will become part of the feature when the feature construction is complete and the datum disappears.

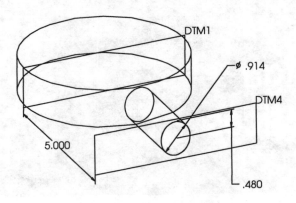

Figure 4-1. A protrusion sketched on a datum on the fly.

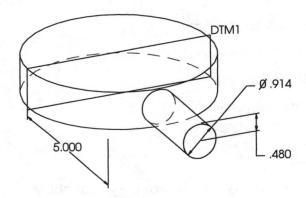

Figure 4-2. The offset dimension to the datum has been internalized.

When Should Datums on the Fly Be Used?

When creating a datum the question should be posed, "Will this datum ever be used again in this part?" If the answer is no, then the datum should be created as datum on the fly. If later, for some reason, the datum is needed, a new datum can always be created with the same constraints. Generally, any feature requiring a sketching surface will give you the option of creating a datum on the fly.

Following is a list of features that have the ability to "make datum."

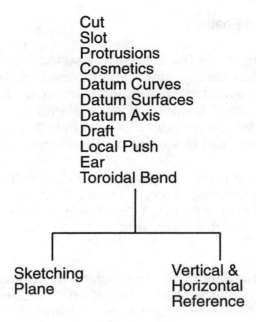

Cut
Slot
Protrusions
Cosmetics
Datum Curves
Datum Surfaces
Datum Axis
Draft
Local Push
Ear
Toroidal Bend

Sketching Plane

Vertical & Horizontal Reference

Figure 4-3.

Attributes for Sketched Features

When creating sketched features, there are certain *feature depth options* available from the SIDES & SPEC TO menus. These are known as attributes and are listed below:

One Side
Both Sides

Blind
Thru Next
Thru All
Thru Until
UpTo Pnt/Vtx
UpTo Curve
UpTo Surface

The choice of which of these attributes to use and when is important in reducing problems that may occur later when modifying the model. To determine which attribute setting best applies to a given situation, review the following discussion of each.

The Notorious Blind Feature

Blind features are perhaps the easiest to visualize and understand. Blind cuts and protrusions simply have a finite depth or height. Though this sounds simple enough, there are hidden problems that can arise when creating Blind features. Usually when a sketched feature is found to have dimensions in the wrong places, this can be fixed by simply redefining the section and changing the dimensions as desired.

However, there is one problem. If the depth dimension of a Blind feature is found to be in the wrong place, it is very hard to change. To illustrate how this problem occurs, Figure 4-4A shows a block with a cutout. Our task is to create a cut in a block and end up with the dimension shown in Figure 4-4A.

A first attempt at making the cutout might be to create a Blind cut on sketched surface A. Figure 4-4B shows what the resulting dimensions would be. Choosing surface B for the same Blind cut would result in the dimension scheme shown in Figure 4-4C. The best way to get the proper dimensioning scheme is to not use the Blind cut at all but to use a *Thru All* cut instead. Figure 4-4D shows how a *Thru-All* cut can be made from datum offset from the back.

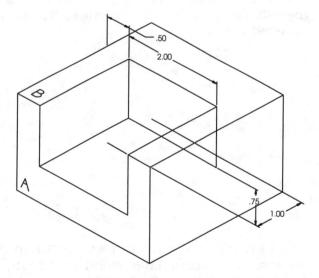

Figure 4-4A.

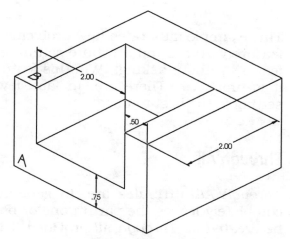

Figure 4-4B.

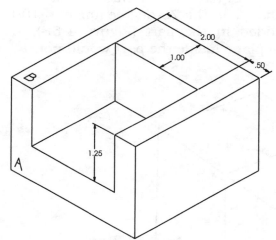

Figure 4-4C.

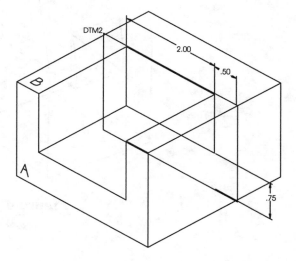

Figure 4-4D.

This example illustrates how problems can arise in dimensioning when a feature with a Blind attribute setting is created. Never construct a Blind feature without asking, "Will this feature be dimensioned correctly when it is finished?" There are, in fact, few times when the Blind attribute setting is the best to use.

Through All

Through All attributes are, in general, best used when a cut or hole completely leaves the part in one or both directions. Any cut or hole can be created as *Through All* provided it leaves the part completely. Figure 4-5 shows why a cut might be made as a *Through All* instead of being created as *Blind*. If the cut in Figure 4-5A (created as *Through All* from DTM1) were made as *Blind*, then a change in the 75-degree angle to 100 degrees would leave the cut embedded in the part (Figure 4-5B). A *Through All* cut created from a datum plane inside the part would not.

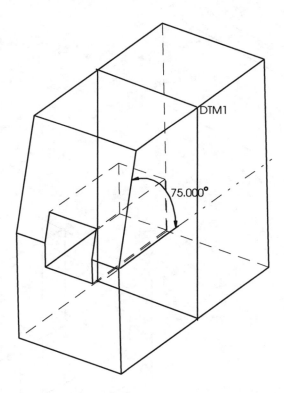

Figure 4-5A.

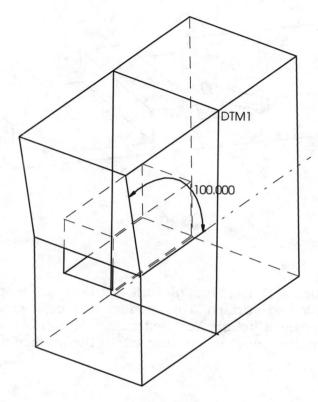

Figure 4-5B.

Through Next

The *Through Next* attribute is used to terminate the extrusion of a feature at the next surface. One restriction applies, however; the entire section of the feature being extruded as *Through Next* must pass through a surface or the feature will not stop.

Figure 4-6 shows two holes created as *Through Next*, both constructed on the same sketching plane. Because one hole passes completely through the curved surface, it stops. The other hole, having not passed completely through the curved surface, continues to the flat surface where it exits completely and stops.

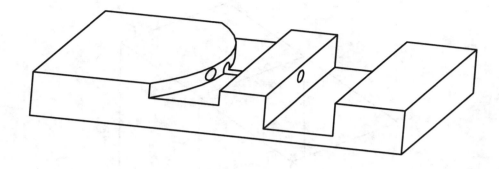

Figure 4-6.

Through Until

Features extruded with the *Through Until* attribute will stop after passing through an indicated surface. This surface can be a solid or a datum plane but, like the *Through Next* extrusion, the entire section of the feature being created must lie entirely within the surface it is terminated on.

Up To Options

The *Up To* options are available for *Extruded* features only (Protrusions, Cuts, Slots, and surface features):

UpTo Pnt/Vtx	Specifies depth up to a plane parallel to the sketching plane and passing through the selected datum point or vertex.
UpTo Curve	Specifies depth up to a plane parallel to the sketching plane and passing through the selected edge, axis, or datum curve.
UpTo Surface	Specifies depth up to a selected surface. For solid features the surface can be another part surface, which need not be planar or it can be a datum plane, which need not be parallel to the sketching plane.

Fill-ins/Cut-off

At times when creating Pro/ENGINEER models it is tempting to fill in an area that was previously cut away. Though this is tempting it should not be done. If the situation arises where a *Fill-in* is desired, consider why the area being filled was ever removed. If necessary, go back to the feature that removed this area and redefine or delete and recreate it. *Fill-ins* are undesirable for several reasons. Not only might they cause machining problems, but they add unnecessary features to the database.

A word of caution and good practice. Do not use *filler* features that negate another feature. An example of such a feature is a protrusion that fills a void, or a cut that removes a protrusion. Make sure that each feature strives to accomplish some goal in the overall design.

Exercise 4-1:   Using some of the techniques just covered, create the block shown in Figure 4-7. Try to construct the model to achieve the dimensioning scheme shown. We want you to see how model construction determines the dimensioning scheme.

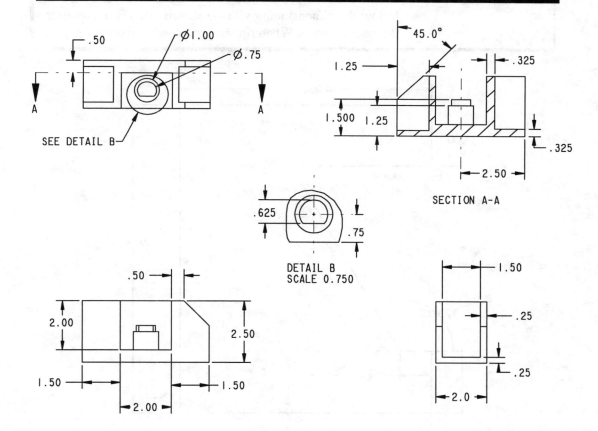

Figure 4-7.

Reference Planes

What does the Vertical and Horizontal Reference do and why is it needed? In Chapter 2 we learned that Sketcher makes assumptions about each line segment in every sketch. One of the major assumptions is that lines that look vertical are vertical and those that look horizontal are horizontal. A logical question then is, "What are these lines assumed to be vertical and horizontal with respect to?" The answer is obviously: "the Vertical and Horizontal Reference."

To fully specify a sketching view, you need to orient the sketching plane to the screen normal axis. You do this by specifying a horizontal or vertical reference plane. This plane must be perpendicular to the sketching plane.

Exercise 4-2:   Create the base feature shown in Figure 4-8. Take care to follow the dimensioning scheme shown. This is a lead-in to the next exercise. When finished, proceed to Exercise 4-3.

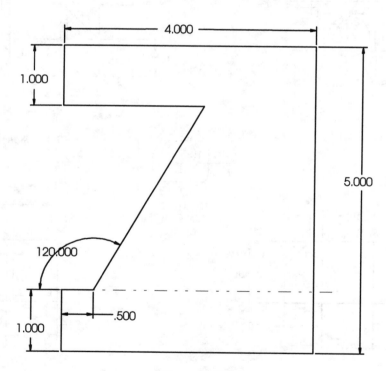

Figure 4-8.

Exercise 4-3:   After the part is extruded, set the 120-degree angle to 90 degrees. Create the protrusion shown in Figure 4-9, being careful to choose the sketching plane and Vertical Reference plane as indicated.

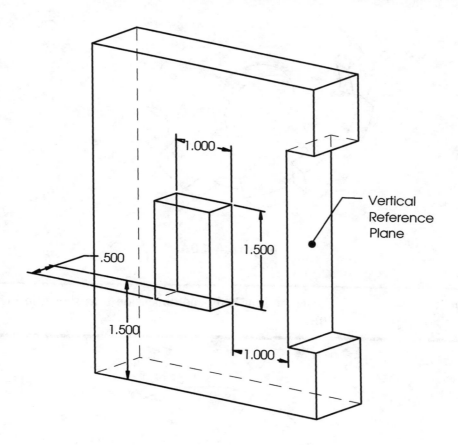

Figure 4-9.

With the second feature finished and extruded, set the 90-degree angle on the base feature back to 120 degrees. What happened to the protrusion? How did this change affect the dimensions?

From this exercise you should see that a feature may be lined up with the Vertical and Horizontal Reference. This is a key concept and deserves much consideration. The following exercise illustrates how we can take advantage of what we just learned and use it to create angular feature patterns.

Exercise 4-4:   Your task is to create the gear shown in Figure 4-10A by patterning the tooth angularly around the perimeter of the gear body.

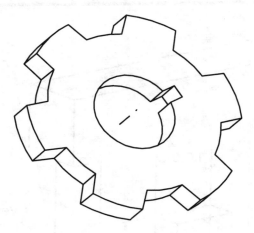

Figure 4-10A.

Step 1:   Create the gear body and dimension it as shown in Figure 4-10B.

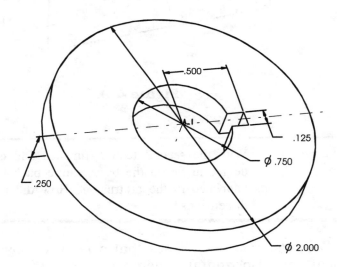

Figure 4-10B.

Step 2: Prepare to create the *Thru All* cut that removes material for the first tooth by choosing the surface shown in Figure 4-10C as the sketching plane. When asked to choose a Vert & Horiz Reference, choose *Make Datum* and constrain the datum "through" the axis and "angled" to the edge of the key-way. See Figure 4-10C. Enter 45 degrees as the angle for the plane.

Questions:

After entering the 45-degree angle, Pro/ENGINEER dropped you into the Sketcher. Why?

_____

_____

_____

Why is the datum on the fly facing the way it is?

_____

_____

_____

Is there another datum shown? If so why?

_____

_____

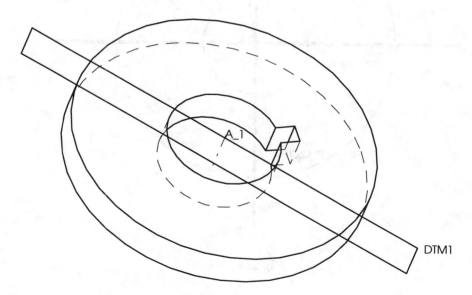

Figure 4-10C.

Step 3:	Construct the cut for the first tooth as shown below.

1.   Under the SEC TOOLS menu, select *Retrieve, Sec, ?*.
2.   Choose *gear_cut.sec* from the enclosed disk.
3.   For the rotating angle, use the default (0.0). When asked to select an origin point for scaling, choose one end of the sketch. When asked to select a point to move, select the other end.
4.   For the scaling factor, use the default (1.0).
5.   When asked to place a section on the part, do so by dragging the section into place.
6.   Now align both ends and regenerate. When the sketch regenerates, choose *Done* and extrude the cut.
7.   Your construction should now look like the figure below.

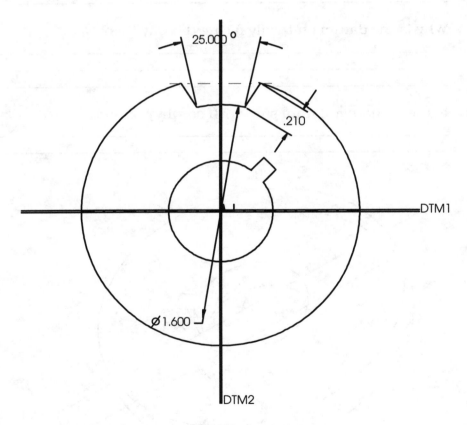

Figure 4-10D.

Step 4: Choose *Modify*, pick the feature, and notice that the 45-degree angle is now associated with the cut. Pattern the cut around the gear six times on 60-degree centers.

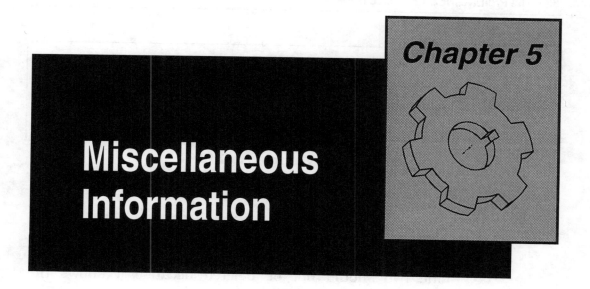

Chapter 5

Miscellaneous Information

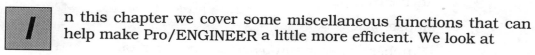

*I*n this chapter we cover some miscellaneous functions that can help make Pro/ENGINEER a little more efficient. We look at

❏ Trail files and creating a generic part

❏ Configuration files

❏ User-defined features (GROUPS)

Trail Files and Creating a Generic Part

A trail file is a record (list) of all menu and screen picks, messages, and keyboard input for a working session. It is an ASCII text file and can be edited as any normal text file. A unique trail file is created every time a Pro/ENGINEER session begins. The trail file, called *trail.txt.#,* is located in the startup directory. The # represents the version number of the file, which the system increments every time a session is started so that old trail files are not overwritten. A sample trail file is shown in Figure 5-1.

```
[1]       !trail file version No. 517
          !Pro/ENGINEER  TM  Release 9.0  (c) 1988-92 by Parametric Technology Corporation  All
Rights Reserved.
          !Select a menu item.
          #PART
          #CREATE
[6]       !Enter Part name [PRT0001] :
[2]       startpart @ ok
          #FEATURE
          #CREATE
          #DATUM
[4]       #PLANE
          #DEFAULT
          #VIEW
```

```
        #FRONT
        !Select front surface.
[3]     0.225319 0.585638 L
[5]     @ sel view 0
        0.869977 -0.288123 -0.228641 0.000000 0.587134 -0.739882
        0.367820 0.681476 0.540786 500.000000 421.875000 -0.000000 1.000000
        0.234840 0.568712 L
```

Figure 5-1. Sample trail file.

Where:

1. Trail file header

   –Line one *must not* be deleted

   –Line two should not be deleted because it shows the release version of the trail file

2. Keyboard input

3. Mouse pick

   –Number shows *x* and *y* screen coordinates

   –Letter shows mouse button. *L* for left, *M* for middle, and *R* for right

4. # represents a menu pick

5. 3D selection

6. ! represents a system message (can be used in documentation as comments).

There are several ways to use trail files. The first and most important use of a trail file is to recover lost work. If the Pro/ENGINEER session is abnormally terminated before your work was saved, the trail file can be used to reconstruct all the work done during that session.

Some Rules to Follow for Editing and Running Trail Files

❑   Trail files must be renamed to something other than *trail.txt* before they can be run. Rename the file to *filename.txt*. This is because another trail file with the name *trail.txt* is created as soon as you start a new Pro/ENGINEER session.

❑   Trail files do not record any use of operating system commands. They do, however, capture changes made to relations tables and family tables in the text editor.

❑   Do not run a trail file that runs another trail file. If a trail file that includes another trail file or that contains operating system commands is to be run, it must first be edited. All lines referring to a secondary trail file or an operating system command must be deleted. Trail files can be edited using any system editor.

❑   The first line of a trail file *must* be the header. These lines cannot be edited out of the file.

❑   The environment of the current session must be the same as that which was present when the trail file was created.

❑   The mode of the current session must correspond to the mode the trail file is expecting, i.e., if the trail file will create features on a part, you must be in part mode.

Note: 1. Be sure to *copy* to new name; do *not* rename.

2. The first pick in trail file must be available.

To run a trail file:

❑   Choose *Misc* from the MAIN menu, then *Trail* from the MISC menu.

❑   Enter the name of the trail file. If the extension is *.txt* it can be omitted, otherwise include the extension.

❑   The trail file will run until it is finished, or until you click on the STOP sign displayed at the right end of the message window.

Another use of a trail file is to build a generic part containing default data and several defined views. When you want to create a new model, run this generic part trail file and you've got your basic part created. If you need to create several similar parts, you can create them all to the point of difference using a trail file. After running the trail file, you can finish the parts by adding their unique features.

Exercise 5-1:   Creating a generic part trail file:

1. Exit and restart Pro/ENGINEER.
2. Create a part made up of the default datum planes (see Figure 5-2).
3. Define and save the following views: Front, Back, Right, Left, Top, and Bottom.

Note: The yellow side of the datum goes to the direction selected in the menu.

4. Select DBMS, Rename, <CR>, then exit Pro/ENGINEER.
5. Pull up a shell window and type *ls*. Note the full name of the latest *trail.txt* file and copy it to a file with a different name by typing *cp trail.txt.<number> startpart.txt*.
6. Edit this file with your UNIX editor. Go to the end of the trail file and delete all lines after: !Enter TO: , and then save the file.
7. Log into Pro/ENGINEER and select *Misc*, *Trail*, and enter the name of your trail file at the prompt.

After the trail file has executed completely, you can type in the new part name when prompted, and you're ready to build your model.

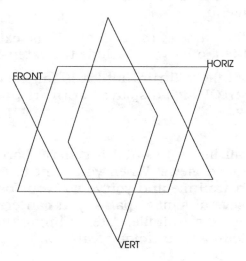

Figure 5-2.

Configuration Files

Configuration files (*config.pro*) are used to preset the environment options and other global settings when starting a Pro/ENGINEER session. These preset options control the way models are displayed, units used, where files are stored, calculation accuracy, tolerance display modes, and so forth.

Pro/ENGINEER can read configuration files from several areas. Each time a new configuration file is loaded, options with the same name are overwritten if new values are assigned. The order and location of configuration files are:

- ❑ **Load point/text:** Your system administrator probably modified the PTC default *config.pro* file to include company standards for formats and libraries. All Pro/ENGINEER users will access this file.

- ❑ **Login directory:** This is the home directory for your login ID. Place your configuration file here so that you can start Pro/ENGINEER from any directory without having a copy of the file in each directory.

- ❑ **Startup directory:** This is the directory you are in when you start Pro/ENGINEER. This directory can be different than your login directory. It may be a special project directory. You control this configuration file.

Configuration files that have the name *config.pro* are read and loaded automatically when Pro/ENGINEER is started. Other configuration files can be loaded, or the *config.pro* file can be reloaded if changes are made to it.

The following pages contain a sample company *config.pro* file with some of the options marked with an asterisk. We will be changing these items in the next exercise. But first, let's review the list.

The asterisks note items to be changed or added in Exercise 5-2.

Config.pro **File Options**

Option	Setting	Remarks
allow_move_attach_in_dtl_move	yes	The *Move* and *Move Attach* commands in drawing mode will act together.
allow_no_geom_storage	no	Environment options *Store Geom* and *No Geometry* are not accessible.
auto_regen_views	yes	Automatically updates objects displayed when you switch between windows.
axis_display	yes	Displays datum axes and their names.
bell	no	Bell will not ring after each prompt.
clock	yes	Displays the clock when Pro/ENGINEER is working.
color_windows	all_windows	Wireframe colors are displayed in the main window and all auxiliary windows.
compress_output_files	no	Output files will not be stored in compressed format.
create_fraction_dim	no	Decimal dimensions will be displayed.
datum_display	yes	Displays datum planes and their names.
default_dec_places	3	Decimal places displayed in all model modes is 3.
dim_offscreen_limit	1.0	Dimensions not allowed outside the screen.
display	hiddenvis	Displays model with hidden lines in gray.
display_in_adding_view	wireframe	Only works when *auto_regen_views* is set to no.
drawing_file_editor	protab	Use Pro/TABLE editor to edit the drawing setup file.
flip_arrow_scale	2.5	Feature creation direction arrow is scaled 2.5 times size.
grid_snap	no	Do not snap to grid. Any location can be picked.

Config.pro File Options (cont'd)

Option	Setting	Remarks
highlight_new_dims	yes	Highlights newly displayed dimensions in red until moved or repaint the screen.
iges_out_spl_crvs_as_126	yes	Converts spline curves to B-spline when creating an IGES file.
iges_zero_view_disp	all_views	Creates a copy of the entity for each view using the view transformation.
info_output_mode	choose	INFO OUTPUT menu will appear. Select: Screen; Write to file; or both.
kbd_cmd_abbreviation	on	Can use abbreviations when entering commands from the keyboard.
mapkey	$f1 #view;#default;#done-return	
mapkey	$f2 #view;#repaint;#done-return	
mapkey	$f3 #view;#names;#iso1;#cosmetic;#shade;#display; #done-return	
✳ mapkey	$f4	
✳ mapkey	$f5	
✳ mapkey	$f6	
!mapkey	$f7	
!mapkey	$f8	
!mapkey	$f9	
✳ mapkey	$f10	
mapkey	$f11 #view;#pan/zoom;#reset;#done-return	
mapkey	$f12 #view;#pan/zoom;#zoom in	
mapkey	/iso1 #view;#names;#iso1;#done-return	
mapkey	/front #view;#names;#front;#done-return	
mapkey	/sh #view;#cosmetic;#shade;#display;#done-return	
mapkey	/top #view;#names;#top;#done-return	
mesh_spline_surf	yes	Turns blue mesh surface lines on.
✳ override_store_back	no	Stores objects in their original directories.
plotter	hp7585a	Establishes the default plotter for plotting files.
pen1_line_weight	4	Sets electrostatic plotter pen weights 1, 2, 3, and 4.
pen2_line_weight	1	
pen3_line_weight	1	
pen4_line_weight	8	
pro_unit_length	unit_inch	Sets the default units to inch.

Config.pro File Options (cont'd)

Option	Setting	Remarks
provide_pick_message_always	yes	Displays message telling what feature or feature's edge has been picked.
repeat_datum_create	yes	Allows repeated creation of datum plane, axis and point. Does not work when using *Make Datum*.
retain_pattern_definition	yes	Retains pattern definition when pattern instances is set to 1.
✳ save_object_in_current	no	Objects retrieved from a "no write permission directory" will not be saved.
save_triangles_flag	no	Will not save the triangles of a shaded view.
shade_surface_feat	yes	Surface features will be shaded.
shade_windows	all_windows	Enables shading of objects in all windows.
show_axis_for_extr_arcs	no	Axes are not created for newly created arcs.
show_dim_sign	no	All dimension values appear positive. Negative values cause geometry to be created to the opposite side.
show_shaded_edges	yes	Shaded edges will be shown darker than the surfaces to which they belong.
sketcher_dec_places	3	Dimensional decimal places in Sketcher is 3.
spin_control	drag	Changes the model orientation continuously by picking on the current value and dragging it along the scale. Picks on scale again to stabilize.
store_modified_draw_models_only	yes	Will only store a model when changes have been made to it (in drawing mode).
store_dependent_objects	changed_only	Only stores dependent objects that have been modified.
tol_mode	nominal	Displays nominal dimension tolerance.

Config.pro File Options (cont'd)

Option	Setting	Remarks
use_dimensioned_edges	yes	Model ridges that are explicitly aligned or dimensioned to will be used to determine the alignment of other Sketcher entities.
x_angle	35	Used to establish a user-defined orientation for models. If these variables are used in the *config. pro* file, the option *User Def* will appear in the ENVIRONMENT menu.
y_angle	35	
* orientation	isometric	Establishes the default view orientation. Depends on *x_angle* and *y_angle* values.

Loading a *Config.pro* File

In general, the *config.pro* file settings are made before you start a Pro/ENGINEER session. If you want to change the environment during the session, it is usually more convenient to use the ENVIRONMENT menu. Some options, however, can only be changed through the configuration file. In this case, you can load a new configuration file during the session.

The new configuration file must reside in the current working directory. It can have any name, which must include the extension and version number, if any.

To load a configuration file during the session:

1. Choose *Misc* from the MAIN menu, then *Load Config* from the MISC menu.

2. Enter the full configuration file name (*config.pro* being the default). Pro/ENGINEER will incorporate changes according to the current configuration file settings. Errors in the new configuration file will be written to the start-up window.

Editing the *Config.pro* File

To edit your configuration file, choose *Edit Config* from the MISC menu and enter the file name (*config.pro* is the default). The Pro/TABLE® editor window comes up with your current settings. You can edit them as desired.

All normal Pro/TABLE commands are available, including the F3 key to enter cell locations to move to (even though row and column numbers are not visible). In addition, you have the use of the F4 (HELP) function key, which does the following:

❑   While in the left (keywords) column, open a subwindow listing all available options. You can select a keyword to place in your *config.pro* file by highlighting it and hitting <CR>, or quit by hitting <Esc>. Entering a letter at the command line will page to the keywords that start with that letter.

❑   While in the right (values) column, all available values for the current keyword will appear in the subwindow.

Notes:

1.   If the value required for the current keyword is a number, or user-defined string (like for "mapkey"), the help subwindow will be empty.

2.   After the editing is completed, load the file using the *Load Config* option in the MISC menu to enable new settings.

Exercise 5-2:   Modify the *config.pro* file using the following options. We will be using the *config.pro* file found on the companion disk.

---

1.   From the MAIN menu choose *Misc.*
2.   From the MISC menu choose *Edit Config.*
3.   Using the F4 Help function key within the Pro/TABLE editor make the following changes and/or additions:

mapkey	$f4 #dbms;#store;#done-return
mapkey	$f5 #regenerate;#automatic;#done
mapkey	$f6 #feature;#create;#datum;#plane;#done;#default
mapkey	$f10 #view;#pan/zoom;#zoom out;#done-return

override_store_back	yes
save_object_in_current	yes
orientation	trimetric

Exit the Pro/TABLE editor and *Load* the *config.pro* file.

User-Defined Features (Groups)

Pro/ENGINEER provides the tools to quickly copy features within the same part, or to copy features to any number of different parts. Two of these tools are *User-defined features (Groups)* and *Local Groups*.

User-Defined Features Tool (Groups)

This tool allows you to place features in a group for duplication in any part. The group of features can be defined in such a way that the duplicated features may look exactly like, or may look similar to but vary in size from, the original features.

Groups can also be table driven. This lets you create a library of specific variations of a group by storing feature dimensions in a table along with an instance name. Each instance will have its own set of values, but will share the reference part, relations, and placement references of the original features.

Local Groups

This tool lets any number of features be grouped together for quick duplication within the active part. The features are duplicated by making a pattern of the group. Once duplicated, the local groups can be unpatterned and ungrouped to act as single features again.

We discuss the creation of local groups in the next chapter.

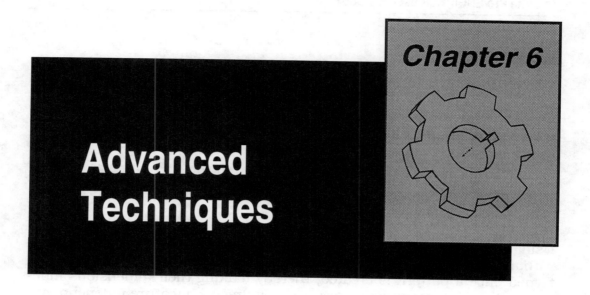

Chapter 6

Advanced Techniques

n this section we cover the following topics:

- ❏ Creating custom feature groups
- ❏ Local groups
- ❏ Creating feature patterns
- ❏ Creating part families

Creating Custom Feature Groups

To create your own user-defined features:

1. First create the feature or features. The features you group must be consecutive feature numbers.

2. Choose *Feature, Group, Define*, and type in a name for the group.

3. Pick all the features you want included in the group, then *Done*.

4. Select *Table* or *NoTable* from the DEFINE GROUP menu.

5. Select all varying dimensions whose values are to be entered by the user when the group is being placed, and enter the prompt text.

6. Next, select and name each external reference surface or edge that was used to place the group and dimension the feature.

7. Answer the prompt to store the group with a reference part.

The message "Storing group information" is displayed followed by "Group 'group name' has been stored." Groups are stored according to the convention *groupname.gph*.

Local Groups

Another custom feature group available in Pro/ENGINEER is the *local group*. Local groups differ from user-defined features in two ways:

❐   Local groups can only be used in the current part.

❐   When creating a local group, you do not give placement references.

Using local groups is a quick way to collect features to copy or pattern without having to specify references. In addition, they can be unpatterned once a pattern is created, thereby making their dimensions independent; and they can be ungrouped, making each feature independent.

To create a local group:

1.   Create the feature or features. They must have consecutive feature numbers.
2.   From the FEATURE menu, choose *Group, Local Group*.
3.   Type in a group name.
4.   Select the features you want grouped together, then choose *Done*.

Creating Feature Patterns

Features can be patterned in a variety of ways. Though some patterning techniques are easier and faster than others, the final decision of which technique to use should depend on the desired dimensioning scheme. In this section, a pattern refers to any two or more features that vary in size or location, but not in shape. There are many ways to create patterned features, but we will concentrate on five major ones:

❐   Patterns created using the *Pattern* command

❐   Patterns created in Sketcher

❐   Patterns created as local groups

❐   Patterns created by copying features

❐   Table-driven patterns

All of these techniques are useful in creating patterned features, but each has different advantages in producing specific types of dimensioning schemes.

Patterns Created Using the Pattern Command

Perhaps the fastest way to create a feature pattern is with the *Pattern* command. This command is useful in creating regularly spaced feature patterns with a variety of options. However, though this method is fast, the resulting dimensioning scheme is rather abbreviated and requires either the addition of driven dimensions or notes to completely dimension the part for manufacturing.

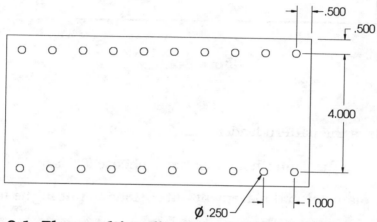

Figure 6-1. The resulting dimensioning scheme for a hole pattern using the *Pattern* command.

Patterns Created in a Table

Patterns can be driven by a pattern table. This functionality allows for the following:

☐ Uniquely locating and sizing each instance of a pattern based on the values for it in the pattern table.

☐ Modifying the pattern simply by creating or reading in another table, selecting dimensions of the pattern features on the screen, or editing the current table values.

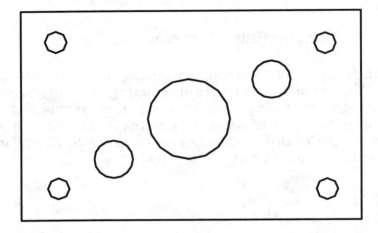

Figure 6-2.

The large hole is the pattern leader.

1. After selecting the center hole to pattern, choose *Table*.

2. Select the diameter and placement dimensions to put in the table.

3. Edit the table and add all the instances and their locations. Note the leader feature will not show in the table.

Patterns Created in Sketcher

In special cases patterns can be created in Sketcher mode. By sketching geometry and mirroring it over and over, a pattern of geometry can be created with no restrictions on the dimensioning scheme.

There are several drawbacks to this method. First, because all the geometry is created on a single sketch, it all has the same depth. Therefore, you could not create a pattern of counterbored holes in this manner. Second, if there are a large number of features in the pattern, the sketch of the features and their dimensions can become very busy and confusing to work with later. Third, some downstream processes, such as Pro/MANUFACTURE, look for specific features in a Pro/ENGINEER database. If such a downstream process is looking for holes and the hole pattern was created as a sketched cut, those features will not be found.

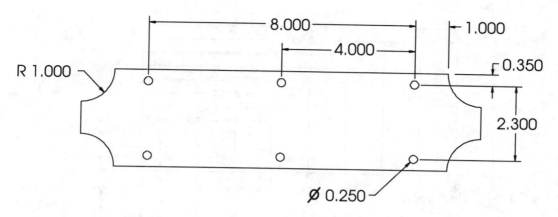

Figure 6-3. Pattern created in Sketcher mode.

Figure 6-3 shows a nameplate with a pattern of six holes, all created as one sketch. Notice that all the holes are dimensioned off of one hole. This could not have been done using the *Pattern* command.

Patterns Created Using Local Groups

For feature patterns in which all features are dimensioned back to a common edge, axis, or datum, the *local group* command can greatly reduce creation time. Normal single feature patterns cannot be unpatterned, i.e., pattern members cannot be dissociated from each other. This isn't true for group patterns. When a feature is unpatterned from a local group pattern, the feature's dimensions are restored. If a local group is created and patterned, then each member is unpatterned, a dimensioning scheme that will look like the one shown in Figure 6-4.

The following steps were taken to create this pattern. Although this may seem like a time-consuming method, it is relatively quick for large patterns.

1.   Create the first member in the pattern.
2.   Create a local group out of the first member. In the example, the first member is a single hole.
3.   Pattern the group using the *Group Pattern* command.
4.   Unpattern each member in the group.

The resulting pattern will be one in which all of the features created in the group pattern will be dimensioned back to the same edge, axis, or datum.

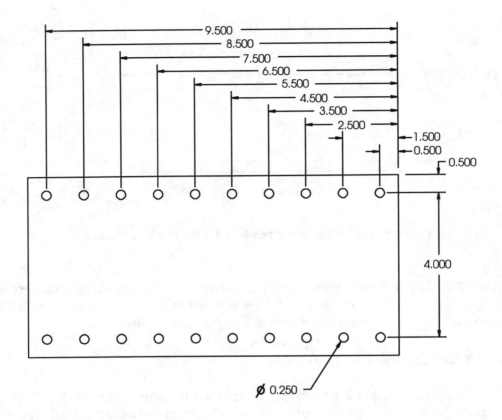

Figure 6-4. The resulting dimensioning scheme for a hole pattern using *Local Group Pattern* command.

Patterns Created by Copying Features

To create a pattern in which only the pattern members at the extent of the pattern actually show their dimensions, the *Copy* option can be used.

One example is shown in Figure 6-5. The desired goal was to create a hole pattern in which the holes in the left side of the part lay on the centerline of the holes in the right side. Also, all of the holes in the part should be dimensioned back to the first hole in the pattern (hole A).

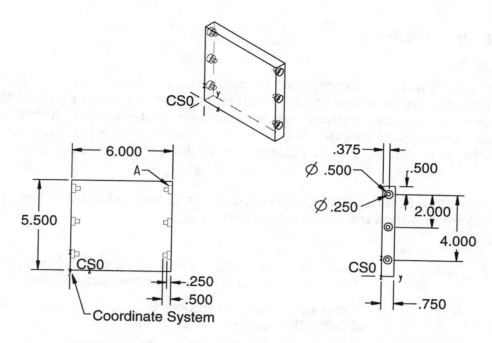

Figure 6-5. Patterns created by copying features.

The following steps were used to create these holes:

1. Create the block as shown.

2. Create a coordinate system as shown.

3. Create hole A as a blind hole along with a coaxial counterbore, using the top edge and the front surface edge for dimension references.

4. Copy hole A, including the counterbore, using *Translate*, picking the coordinate system, and entering the correct displacement.

5. Repeat step 4 for the third hole.

6. Now copy the three features to the other end of the plate by *Reference (New Refs)*, using the same dimensions and selecting a new placement surface and new reference edges. Select the same coordinate system.

7. Display dimensions to check dimensioning scheme.

Table-Driven Patterns

Features can be patterned by using pattern tables. Pattern tables simplify the creation and modification of patterns, by controlling all the dimensions for the pattern instances. Each pattern table drives its pattern throughout its existence. Multiple tables can be established for a pattern. You can change the pattern by switching the table that drives it.

Pattern table functionality is accessed through the PATT TABLE menu. The menu contains the following options:

Edit	Brings up an editor window (default part table editor) to modify the content of the pattern table.
Add	Brings up an editor window to create a new pattern table for the current feature.
Remove	Deletes a pattern table from the current feature.
Rename	Changes the name of a pattern table for the current feature.
Switch	Makes a different pattern table drive the pattern for the current feature.
Write	Outputs the current pattern table to an ASCII file with the extension *.ptb*.
Read	Imports a pattern table from an ASCII file with the extension *.ptb*.
Done	Completes changes to the pattern table for the current feature.
Quit	Aborts changes to the pattern table for the current feature.

Modifying Table-Driven Patterns

There are three ways to modify a pattern that is table driven.

❒ You can modify any pattern dimension by picking the dimension, entering a new value, and regenerating the model. Modifying values on the screen causes them to be modified in the table as well.

❏   You can modify the pattern by editing the table directly. This allows you to modify multiple values for multiple instances and regenerate all of the changes at once.

❏   You can replace the current table with another table, causing the pattern to be driven by a completely different set of values.

Exercise 6-1:   Create the part shown in Figure 6-6.

1.   Begin with the default datum planes.

2.   Create the block using the dimensions shown. Position the block so that two datum planes pass through the position of the first hole (lower left-hand corner).

3.   Now create the hole. Select *Feature, Create, Solid Hole, Dim Pattern, Linear, Done*. Select *Thru All* and enter the diameter.

4.   When prompted to select the placement plane, select a point above and to the right of the two datum planes. Select the two datums for dimensioning and accept the default dimensions.

5.   Select *General* and *Done*. From the PAT DIM INER menu, select *Table*, then select the two dimensions from step 4. *Done*. From the PATT TABLE menu, select *Add* and enter the name for the pattern table.

6.   When the Pro/TABLE displays, type in the following values.
Table Name XXXXXXX

idx	d1	d2	(note your particular
1	6.25	0	dimension's orientation)
2	12.50	0	
3	0	3.25	
4	12.50	3.25	
5	0	6.5	
6	6.25	6.5	
7	12.5	6.5	

7.   Exit the Pro/TABLE editor and select *Done* and *Done*.

8.   Now modify your first hole's dimensions back to zero. Store the part for later use.

Do you get the dimensioning scheme shown in Figure 6-6?

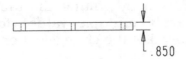

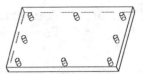

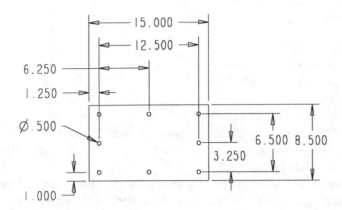

Figure 6-6.

Creating Part Families

Like the standardized parts you buy from a catalog, families of parts (also called *table-driven* parts) are collections of similar parts that are available in different sizes or have slightly different detailing features. For example, screws come in all different sizes, but they all perform the same function and look somewhat alike. Thus, it is convenient to think of them as a family of parts.

Table-driven families provide a very simple and compact way of creating and storing large numbers of objects. In addition, family tables promote the use of standardized components and allow you to represent your actual part inventory on Pro/ENGINEER. Moreover, families facilitate interchangeability of parts and subassemblies in an assembly; instances from the same family are automatically interchangeable.

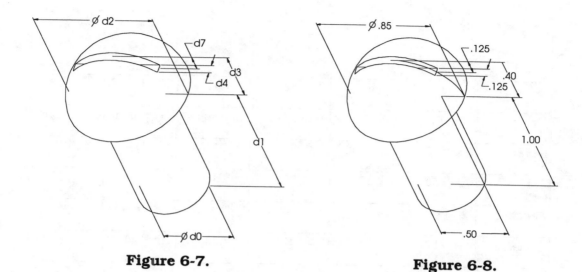

Figure 6-7.   **Figure 6-8.**

Thus, this generic screw can be created along with a table of dimensions that will define our family. Both symbolic and actual dimensions are shown in Figures 6-7 and 6-8.

Exercise 6-2:

1. Create the generic screw following the dimensioning scheme shown in Figures 6-7 and 6-8.
2. From the PART menu, select *Family Tab, Add Item, Dimension,* select the dimensions in order, then select *Done/Return.*
3. To include descriptive names of the symbolic dimensions in the table, choose *Modify, Dim Cosmetics, Symbol* and pick each dimension. Type in the symbolic name for that dimension. See the table in Figure 6-9. Continue naming all dimensions as shown in the table. Select *Done, Done.*
4. Go to the FAMILY TABLE menu, select *Edit,* and add the information shown in Figure 6-9 to your Generic Screw Table. Be careful when typing in data.
5. Exit the Pro/TABLE Editor.

	d0	d1	d2	d3	d4	d7
	thd_dia	thd_len	head_dia	head_ht	slot_wid	slot_dep
Generic	.500	1.000	.850	.400	.125	.125
Screw_1	.480	.987	.800	.380	.120	.120
Screw_2	*	*	*	*	*	*
Screw_3	.250	.750	.420	.200	.120	.120

Figure 6-9.

Exercise 6-2 (cont'd):

6.   To view an instance, choose *Instance* and select an instance listed in the Pop-up menu. *Quit Window* to exit the small window. *Change Window* and select on the large graphic window to make it active. Store the part and *Quit Window*.

7.   In the assembly mode, create an assembly and retrieve the block of Figure 6-6. Then assemble instance Screw_3 of the generic screw part into one of the holes. Too small? Replace it with one the proper size.

8.   Pattern the screw. Select *Pattern* from the COMPONENT menu, pick the screw, and select *Ref Pattern*, and *Done*. The screw will be placed in the hole pattern. If you selected *Ref Pattern* in step 7, there is no need to do so for step 8.

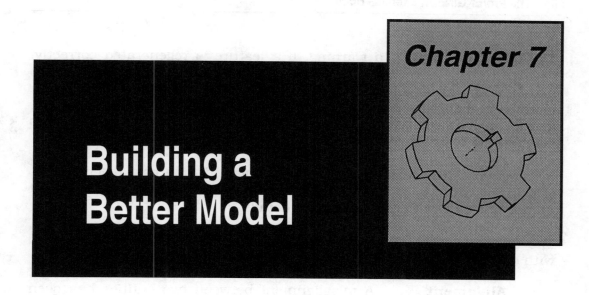

Building a Better Model

Chapter 7

Building a model that meets all of the concurrent engineering requirements of analysis, drafting, prototyping, and manufacturing is important. If a model does not meet these requirements, then it is not a complete model.

In this section we examine the following:

- ❑ Geometry checks
- ❑ Driven dimensions
- ❑ Stereolithography apparatus files
- ❑ Manufacturing requirements
- ❑ Tolerancing
- ❑ Accuracy
- ❑ User tips
- ❑ Deciphering error messages and warnings

Geometry Checks

To prevent regeneration problems, geometry is checked for possible geometry errors by Pro/ENGINEER. Using the *Geom Check* option in the INFO, INFO REGEN, and TRIM PART menus, you can view the feature that Pro/ENGINEER says has a possible error, then work to correct it. Normally the TRIM PART menu is dimmed out. If Pro/ENGINEER detects geometry that may cause regeneration problems, this option becomes available.

The geometry check is a warning. If a feature is regenerated correctly, then you might never have a problem. However, the warning is there to highlight an area where you might eventually have a problem after modifying the part, and it allows you to revise the definition of the feature to correct any problem. Not correcting a geometry check is comparable to driving your car with the oil light on.

If Pro/ENGINEER detects a potential problem during part regeneration, a warning will be issued to alert the user: "Warning: misalignment found. Use GEOM CHECK menu for details."

The following types of geometry errors may occur and can be checked with options in the ERR_GEOM menu:

Alignment   A misalignment between two entities has been detected. This is usually caused by very small edges, resulting from dimension modification of the part. You are being warned that perhaps you should redefine the feature to align explicitly, or to increase accuracy so that the small edge does not cause regeneration problems later. A good rule of thumb is that the ratio of the shortest element to the longest element should not be less than about 0.0012.

Overlap   An overlap of one feature on another has been detected. This generally occurs when retrieving versions of parts last stored in an older release of Pro/ENGINEER. Pro/ENGINEER has improved geometry checking to where overlapping geometry will not be allowed when you try to create the feature.

Unattached   This can occur when a feature has become unattached, such as an open section feature that has been moved inside the part or a hole that has been moved off the solid into space.

Round Errs   When a round cannot be created anymore, this error diagnoses the problem. This usually occurs when the part has been modified and an edge that was rounded no longer exists, or when it is impossible to make a round on one of the selected edges with the radius specified.

Incomp Loop   This can occur wherever a tangent edge of a round is being partially intersected by another feature's edge.

The example in Figure 7-1 shows a couple of situations that could cause a geometry check. The round, revolved cut, and protrusion were all created on the base feature. Assuming that the features were not aligned with each other, a geometry check could occur at the two intersections between the features.

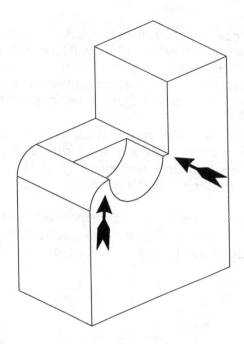

Figure 7-1.

There are two ways to fix a geometry check warning. If the design intent is for all of the features to intersect each other, then you will have to go back and redefine each feature to explicitly align with the previous feature.

If the intent is not to align the features, then the accuracy of the part can be increased to eliminate the geometry check warning. As a general rule the accuracy should be set to equal the ratio of the smallest dimension on the part over the largest dimension.

Avoid using blind features to extrude all the way through a part or intersect another surface. If the blind depth is not equal to the part, a gap will be created. Pro/ENGINEER may pick this up and issue a geometry check warning. To correct this situation, redefine the feature and use one of the *through* options.

Sketching features on drafted surfaces may also cause a geometry check warning. If this is desired, use the use *Edge* command to align the entities.

In order to avoid future problems with your model and make it as clean as possible, all geometry checks *must* be eliminated from your model.

Driven Dimensions

Driven dimensions are created in Drawing mode. They measure the model geometry on which they are placed. Driven dimensions have one-way associativity. They cannot be modified, but they will reflect any changes to the model in part or assembly mode. Because driven dimensions are not two-way associative, they should be used with care.

Why would we need to add more dimensions to a drawing when Pro/ENGINEER will display all the model dimensions on the drawing? The only reason to create a driven dimension is that the model does not contain the required dimensioning scheme.

In the model in Figure 7-2 and the corresponding drawing in Figure 7-3, notice that the dimensions on the drawing are not the same as the model dimensions. The "200" and "300" dimensions in Figure 7-3 are the driven dimensions.

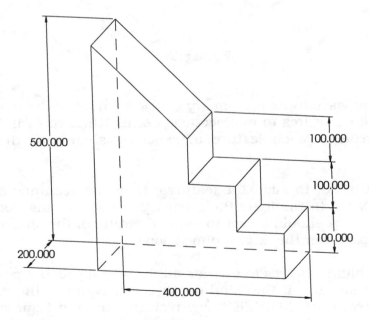

Figure 7-2.

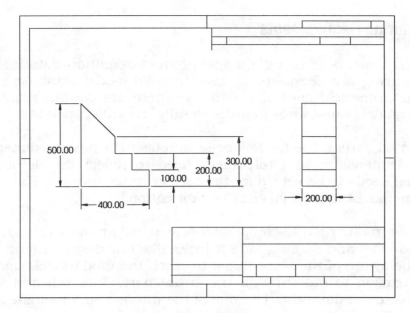

Figure 7-3.

As stated earlier, one way to avoid having to use driven dimensions is to build the model so that it contains all the dimensions that you want to show on the drawing. In this example we could redefine the cut to reflect the desired dimensioning scheme.

To avoid drafting problems and to model the design intent of the part, driven dimensions should be eliminated from the drawing.

Stereolithography Apparatus Files

Stereolithography apparatus (SLA) is a new technology that allows the user to produce a physical prototype (made from a liquid resin) in a significantly cheaper and shorter cycle time than traditional methods. Applications and benefits include enhanced visualization, form, fit, and function testing, cable routing, assembly and tooling methods, and the generation of patterns for investment castings.

The SLA technology, as well as all current rapid prototyping technologies, uses a data format called an STL file to transition between the solid model and the rapid prototyping technology hardware. Before interfacing with the rapid prototyping laboratories, it is important to generate a test STL file to ensure that the PTC database is complete and without any geometry discrepancies.

Manufacturing Requirements

One of the benefits of using a single product definition database is that manufacturing is able to directly use the solid model database for many computer numerical control machines. There are certain requirements that our models must meet in order to fully use this capability.

As we stated earlier, the models need to reflect the desired dimensioning scheme. Dimensions and tolerances need to reflect the design intent. The model needs to be set up for the correct units and material. This will determine machining speeds during fabrication.

Sometimes in our rush to complete a design, we are only concerned with the final model and making sure it looks like our design intent. We also need to be aware of the steps taken to reach the final model. Don't use a blind extrusion to cut through the entire part. This will add an extra dimension to the database. Use one of the *through* options instead. Avoid the use of fill-ins. If you need to add material, why was it removed in the first place?

Above all, ensure that there are no errors in the model, because this information may be passed directly to the manufacturer.

Tolerancing

When a model is completed, the final step is to put the correct tolerances on the model. Pro/ENGINEER will put a default tolerance on all dimensions of a model. To view these tolerances in model mode, the *Tolerance* option in the ENVIRONMENT menu must be set to *On*. The tolerance can be displayed in several formats: Nominal, Limits (maximum and minimum values), Plus-Minus (+0.010 -0.005), and ± Symmetric (± 0.010). The format can be changed by using the *Modify*, FORMAT menu options. When the tolerance is displayed, it can be modified just like any other dimensional value.

Accuracy

You must remember to incorporate correct modeling practices into the models. If you do not get rid of geometry checks, you may develop problems in the future when redefining models. You may want to avoid driven dimensions because they are not fully associative. If you need them on the drawing, you should redefine the model to reflect the design intent. Most of all, you need to realize that other groups such as manufacturing and analysis will be using the model and they require correct geometry, tolerances, dimensions, and no errors.

The following checklist can be used as a reference to determine if the model meets all concurrent engineering requirements.

Checklist

Pass Fail
Yes No
[ ] [ ]   The model does not contain geometry checks.
[ ] [ ]   The drawing will not have driven dimensions.
[ ] [ ]   An STL file can be generated from the model.
[ ] [ ]   Dimensions and tolerances are correct.
[ ] [ ]   Correct units and materials are specified.
[ ] [ ]   Correct geometry is modeled.
[ ] [ ]   There are no errors in the model.

Exercise 7-1:	The part shown on the next page was designed and modeled using poor modeling practices. Management wants it corrected because it is causing problems in manufacturing. You are given the job of fixing what someone else modeled without knowing how he or she did it. Follow the tasks outlined.

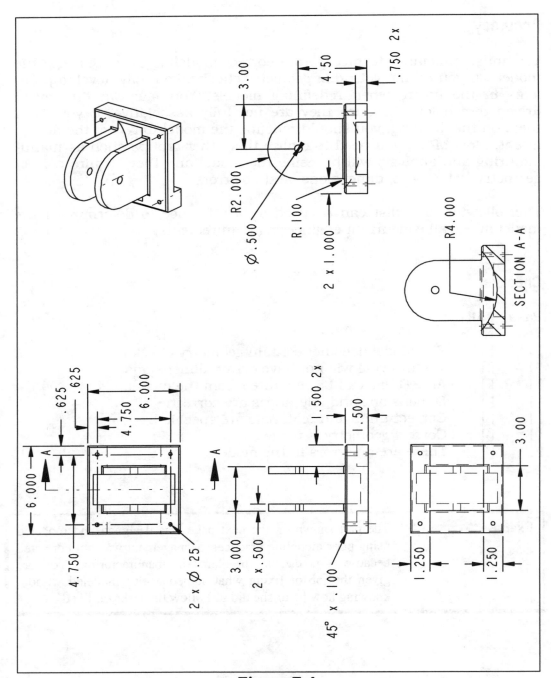

Figure 7-4.

Note: There may be a geometry check in this exercise. This is only a warning, but if one is found, it must be corrected.

Task 1:   Modify the part to remedy the driven dimensions (the 4.50-inch dimension in the right drawing view and the 1.500-inch dimension in the front view) and make them feature or model dimensions.

Task 2:   There are nine drilled holes in the base feature. Five of these are plugged. Correct this situation so only the four corner holes remain.

Task 3:   Modify the 3-inch dimension between the wheel support to 4 inches, while keeping the protrusion centered on the base feature.

Task 4:   Modify the tolerance of the base feature, 6" x 6" x 1.5", to two decimal places. Experiment with the other tolerance options.

Task 5:   Make the 4-hole location dimensions basic dimensions and add a geometric tolerance to the 0.25 diameter hole location. Make this a true position of 0.010 inch relative to datums TOP, FRONT, and LEFT.

User Tips

These user tips and error messages and warnings are provided to give some general guidelines for the new user to follow.

- ❏   Always start with datums as the base feature.
- ❏   Name all features.
- ❏   Set and name orientation views.
- ❏   Envision the machining processes and sequences as you create each part.
- ❏   Put a drawing in the subwindow, and dimension features as you create them; redefine dimensioning scheme if necessary.
- ❏   Use local groups to create hole patterns so the dimensioning scheme is correct and may easily be modified.
- ❏   Create models of parts with pins and so forth as assemblies, but create the drawing as an inseparable assembly.
- ❏   Always *Align* to surfaces, not edges (use ISO view).
- ❏   Put rounds and chamfers on last.
- ❏   Use surfaces rather than edges during part orientation and feature dimensioning.
- ❏   Store at least twice a day.
- ❏   Get help from the following sources:
  - –   Read the manual
  - –   Take training classes

Deciphering Error Messages and Warnings

- ❏   Bad expression on a line: look at the relations file.
- ❏   Frozen part indicates
  - –   Missing layout or missing part files
  - –   Missing assembly constraints
- ❏   Unattached feature: often caused by cross-sections that are open; also caused by putting features in "space."
- ❏   First attempt at part regen failed: invalid dimensions.

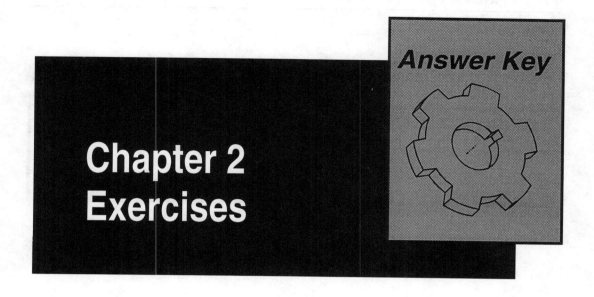

Chapter 2
Exercises

Answer Key for Exercise 2-1

Exercise 2-1: Using the information just covered, fill out the feature form, Figure 2-4, for the cut on the part in Figure 2-3.

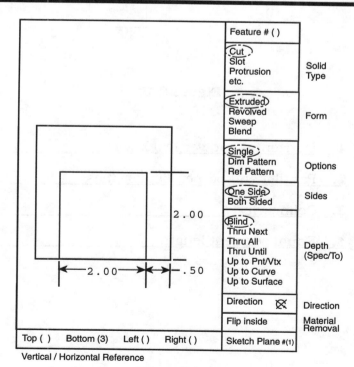

Figure 2-4.

Answer Key for Exercise 2-2

Exercise 2-2:	Using the information in Table 2-1, sketch the following sections and try to determine the Sketcher assumptions made.

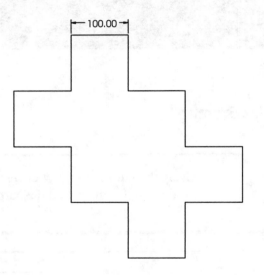

Figure 2-6.

1.   Horizontal and vertical lines

2.   Parallel and perpendicular lines

3.   Colinearity

4.   Equal segment length

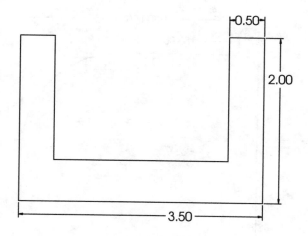

Figure 2-7.

1. Horizontal and vertical lines

2. Parallel and perpendicular lines

3. Colinearity

4. Equal segment length

Note: The assumption that the GAP around the figure is the same is not made by Sketcher.

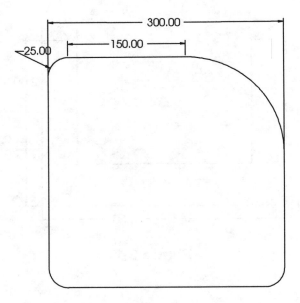

Figure 2-8.

1. Equal radius/diameter

2. Horizontal and vertical lines

3. Parallel and perpendicular lines

4. Tangency

5. Equal segment length

Answer Key for Exercise 2-3

Exercise 2-3:   Accuracy problems can be overcome in many instances by sketching the section using exaggerated dimensions, regenerating, and gradually approaching the actual dimension. Try this on the following exercise. Create and regenerate the sketch shown in Figure 2-10. Create all necessary construction geometry.

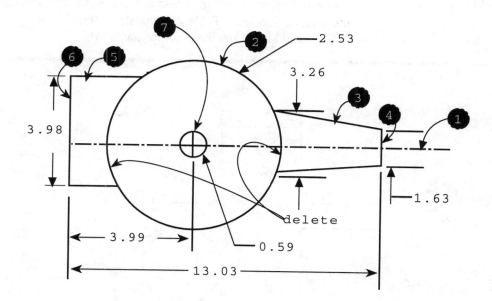

1.   Sketch a horizontal center line.

2.   Sketch a circle with its center on the centerline.

3.-6.   Sketch half of the pointer, exaggerating its proportions and angles as shown.

7.   Sketch a smaller circle with its center coincident with the large circle.

8.   Mirror items 3 to 6 about the centerline. Do not include the two circles.

9.   Use *Geom Tools* from the SKETCHER menu, select *Intersect*, and pick the large circle near one of the intersecting lines. Then pick the line. Repeat this operation for the remaining three intersecting locations on the large circle.

10.   *Delete* the two arc segments shown. *Done Sel.*

11.   Dimension the sketch as shown. **Note:** the dimension values shown above were the sketch defaults.

12.   *Regenerate, Done, Modify* change the dimensions to the values in Figure 2-10.

Answer Key for Exercise 2-4

Exercise 2-4:	Using the information on points and centerlines above, and the information from Table 2-1, construct the sketch shown in Figure 2-11 and list the assumptions used.

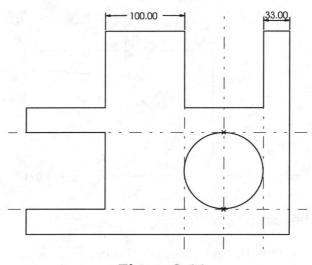

Figure 2-11.

1.   Horizontal and vertical lines

2.   Parallel and perpendicular lines

3.   Tangency

4.   Colinearity

5.   Equal segment length

6.   Point entities lying on other entities

Exercise 2-5:   Create and regenerate the sketches shown in Figures 2-12 and 2-13. Create all necessary construction geometry. Remember the assumptions and rules that were just covered (especially points).

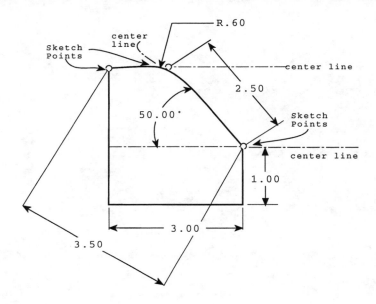

Figure 2-12.

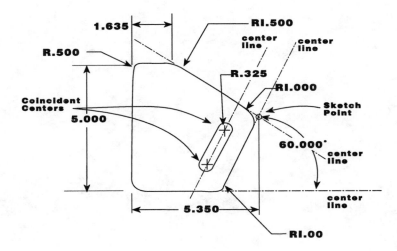

Figure 2-13.

Chapter 3
Exercises

Answer Key

Answer Key for Exercise 3-1

Exercise 3-1:	In this exercise you will reroute the references for the slot feature located in the lower left of the part shown in Figure 3-1.

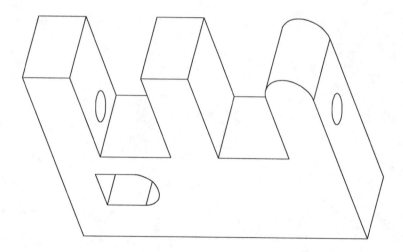

Figure 3-1.

Select *feature.prt* on the disk.

You want to reroute the sketching plane from the front surface to the back surface. Reroute the orientation plane from the top of the middle post to the top of the left post. Reroute the vertical locating dimension from the front edge of the middle post to the front edge of the left post.

Start by getting the regeneration information.
Select *Info* from the PART menu, *Regen Info* from the INFO menu, then select *Beginning* to start the part regeneration from the first feature.

Now reroute the slot. The following are the menu picks to accomplish the reroute.
>   *Feature/Reroute*

Select the slot feature, answer *yes* to the prompt to roll back the part.
>   *Query Sel*

Select the back surface.
Next, select the top of the left post as the alternate horizontal reference.
Next, select *Same Ref* for the horizontal dimension reference, then select the front edge of the left post as an alternate for the vertical dimension reference.

At this point the reroute is complete, but the slot has disappeared. This happens because the sketching plane is now on the back surface with its direction unchanged, in other words, away from the part.

To change the direction of the cut, select *Redefine* from the FEAT menu, *Sel By Menu, Last Feature, Direction, Done*. Flip the direction of the arrow and the cut reappears. The slot has now been rerouted.

Answer Key for Exercise 3-2

Exercise 3-2:   Using the *Reroute, Reorder,* or *Redefine* commands, make the part *guideblk.prt* look like the following drawing. *Do not delete any features.*

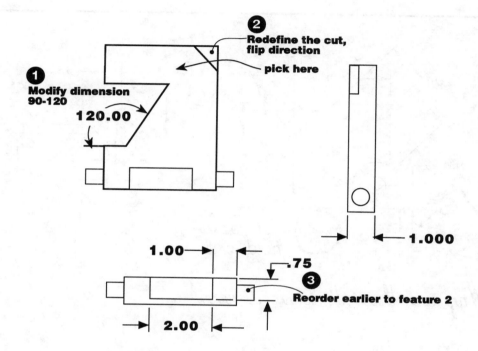

Figure 3-3.

1.   *Modify.* Pick the cut feature, and change 90.00 to 120.00. *Regenerate*

2.   *Feature, Redefine.* Pick the cut as shown.
     Flip, Done
     Flip, Okay. Change the arrow direction.

3.   *Feature, Reorder, Earlier.*
     Pick the round protrusion.
     Answer "*y*" to reorder from feature 3 to 2.

Answer Key for Exercise 3-3

Exercise 3-3:   Make the changes to the part *slavegr.prt* to reflect the changes marked on the following drawing. Use *Regen Info* to get feature numbers. Use the feature numbers when selecting features to work on.

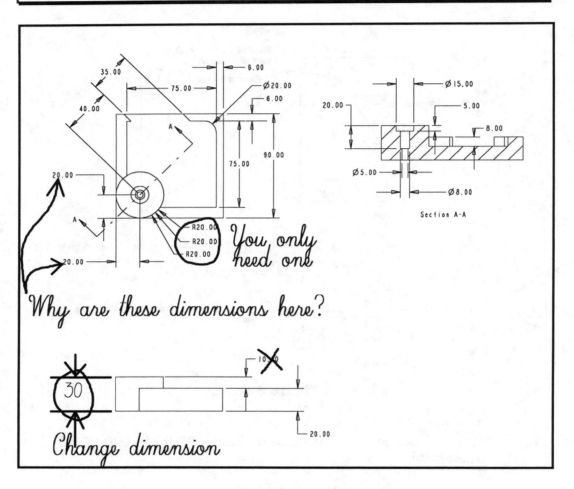

Figure 3-4.

1.   Redefine the protrusion feature 2. Align to the base feature round.

Feature, Redefine. Pick the round protrusion feature 2.
Section, Done
Delete the 20.00 dimension.
Done Sel
Alignment, Align. Pick the sketched circle and the rounded edge of
        the base.
Done Sel
Regenerate, Done

2.   Redefine the cut feature 3. Align section to the protrusion. When
        the protrusion feature 2 is redefined, the center location
        will be known implicitly through the Alignment to the base
        round; therefore, these are extra dimensions and must be
        deleted.

Feature, Redefine. Select the cut feature 3.
Section, Done
Delete the 20.00 diameter dimension at the boss as well as the
        two 20.00 dimensions locating the circles center.
Done Sel
Alignment, Align. Pick the sketch circle and an edge of the round
        protrusion.
Regenerate, Done, Flip, Okay

3.   Reroute the sketching plane to the bottom of the base feature,
        then modify the height to 30 from 10.

Feature, Reroute. Select the round protrusion.
Answer "*y*" to roll the part back.
Select the back surface of the base as the alternate sketching
        plane.
Query Sel, Next, Accept
Same Ref, Same Ref, Same Ref

Answer Key for Exercise 3-4

Exercise 3-4:
1. What will happen if feature 2 is deleted?
2. If we wish to delete feature 3 and not delete feature 11 at the same time, what must we do?
3. What command (*Reroute*, *Reorder,* or *Redefine*) would you use to break the parent/child relationship between features 3 and 11?
4. How far back can feature 9 be reordered?

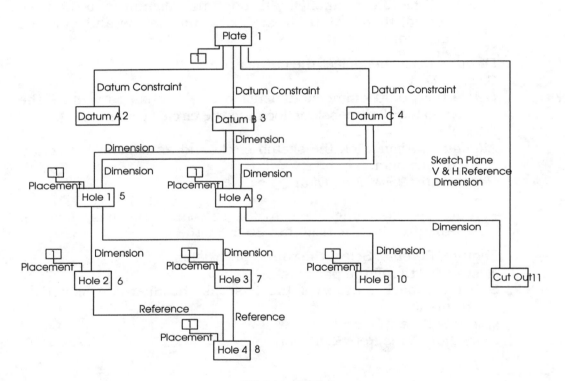

Figure 3-6.

1.   Just Datum A will be deleted.

2.   *Redefine* feature 11 not to reference Hole A.

3.   *Redefine*

4.   To feature 5

Exercise 3-5: Using the space provided and following the steps described in Example 1, create a family tree for the part *vice_bas.prt*, shown in Figure 3-7.

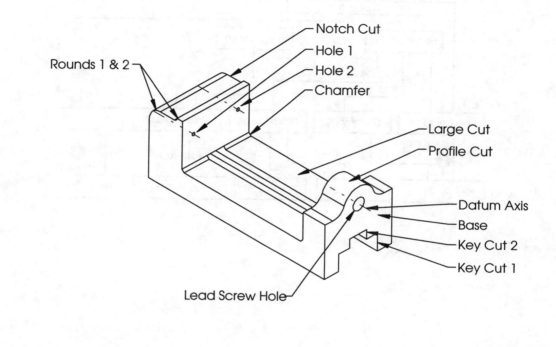

Notch Cut
Hole 1
Hole 2
Chamfer
Rounds 1 & 2
Large Cut
Profile Cut
Datum Axis
Base
Key Cut 2
Key Cut 1
Lead Screw Hole

Figure 3-7.

Use this page to construct the family tree for the *vice_bas.prt*. Start with the base.

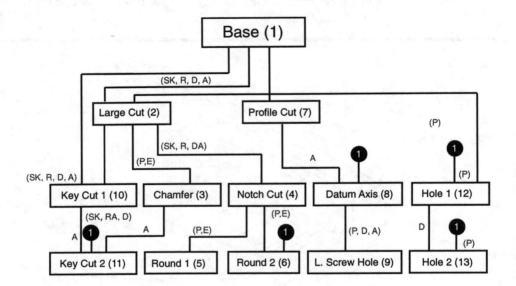

Is this the best possible scheme for this part?   No.   Could you improve on it? Yes.

Exercise 3-6:   Rescheme *vice_bas.prt* such that the parent/child relation-ship tree matches the following one. Mark your corrections on Figure 3-8.

The following needs to be accomplished.
1. Break the Key cut 2 (11)-Chamfer link.
2. Break the Screw hole (9) - Datum Axis (8) link.
3. Break the Hole 1 (12) & Hole 2 (13) - Large cut (2) link.

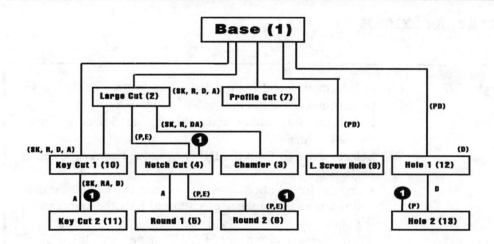

Figure 3-8.

1.   Unalign from Chamfer dimension to base.

2.   Rescheme from coaxial to linear and use the base feature for dimensioning.

3.   Reroute Hole 1. Select two edges not associated with the Large cut for locating the hole.

   When asked to select an alternate surface for the feature placement make a datum offset from the current placement surface a distance of zero (0).

   Reroute Hole 2 to this new datum plane also.

   Redefine the Attributes of Hole 1 and 2 to *Both Sides* and *Thru Next*.

Modify the offset value of the datum plane so that it becomes coincident to the back of the base feature.

Now *Reroute* the two Holes to reference the back of the base feature instead of the datum plane.

The holes no longer reference the Large cut, and no new geometry was created.

Answer Key for Exercise 3-7

Exercise 3-7:   Gather information and build a local family tree for the center protrusion of *link_hol.prt* shown in Figure 3-10.
1. First gather information about this feature. Select *Info* from the PART menu.
2. Select *Feat Info* from the INFO menu and select the center protrusion.
3. From the Information Window, extract the feature number and the note the number of children. Exit the Information Window.
4. Select *Feature* from the PART menu and then select *Suppress* from the FEAT menu.
5. Select the feature to suppress. In this case, select the center protrusion.
6. As each child is highlighted in blue, select *Info* and note the feature number. Let's begin the family tree. See Figure 3-9.
7. Exit the Information Window and select *Suppress*. The next child highlights. Get the information on this child, add the information to the next box of the family tree, and suppress the child.
8. Continue the process until all children have been charted.
9. Quit the *Suppress* function.

1.   *Info* from the PART menu.

2.   *Feat Info.* Select the center protrusion.

3.   The feature number is 6.
The children are 7, 10, 11, 12, 13, 14, 15, 16, 23, 27, 28, 30, and 31.

Type "*q*" to quit the info window.

4. *Feature, Suppress.* Select the center protrusion.
   Info. Type "*q*" to exit the information window.

5.-8.

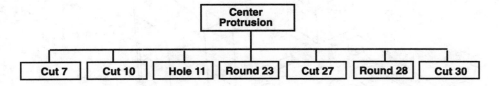

9. *Quit, Quit Del/Sup*

Answer Key to Exercise 3-8

Exercise 3-8: You have just been given the task to modify the part *link_hol.prt*, Figure 3-10, to incorporate two changes listed below. Use the local family tree that we just developed to assist you in this process. *Do not at any time delete any features other than the one listed.*

1. Delete the center most of the three posts.
2. Make the square cut in the posts stop as it passes through the cut in the second post.

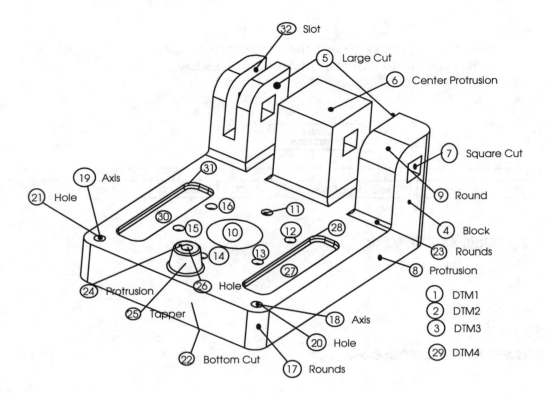

Figure 3-11.

1.   Delete the post:
      Feature, Redefine. Select the round (feature 23).
      References, Remove all edges of the post as referenced.

      Feature, Delete. Select the post.
      Show Ref. Get the references from the CHILD menu.
      Reroute these references rolling back the part to references not on
            the center post.
      Do this for all the children.

2.   *Feature, Reroute.* Select the slot 32. Change the tangent edge
            formed by the round as a reference to the back edge of the
            left post.
      Feature, Reorder, Earlier. Select the slot 32. Reorder to feature 6
            just before the square cut.
      Feature, Redefine. Select the square cut.
      Attributes, From To, Done
      Select surface of the protrusion 8 and *Query Sel* the surface of the
            slot to stop the cut at Figure 3-10.
      Done

Chapter 4
Exercises

Answer Key to Exercise 4-1

Exercise 4-1: Using some of the techniques just covered, create the block shown in Figure 4-7. Try to construct the model to achieve the dimensioning scheme shown. We want you to see how model construction determines the dimensioning scheme.

```
!Select a menu item.
#PART
#CREATE
!Enter Part name [PRT0001]:
4-1
#FEATURE
#CREATE
```

1.50

2.50

#PROTRUSION
#DONE
#DIMENSION
#GEOM TOOLS
#TRIM
!Select a bounding entity to trim to.
#CORNER
#REGENERATE
!Regeneration completed successfully.
!Enter a new value [10.9800]:
2.5
!Enter a new value [12.0000]:
1.5
#REGENERATE
!Regeneration completed successfully.
#DONE
!Enter extrusion DEPTH [0.50]:
2
!FIRST FEATURE has been created successfully.
#DONE
#FEATURE
#CREATE

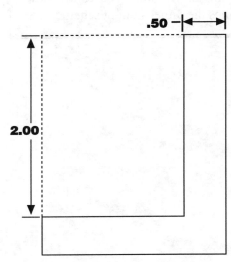

#PROTRUSION
#DONE
#DONE
!Select option for the TO end of the feature.
#DONE
#OKAY
!Select or create a horizontal or vertical reference for sketching.
#TOP
#GEOM TOOLS

#USE EDGE
#SKETCH
#GEOM TOOLS
#TRIM
!Select a bounding entity to trim to.
#CORNER
#REGENERATE
!Regeneration completed successfully.
#DIMENSION
#REGENERATE
!Regeneration completed successfully.
#DONE
!Enter depth [0.88]:
2
!PROTRUSION has been created successfully.
#VIEW
#DEFAULT
#CREATE

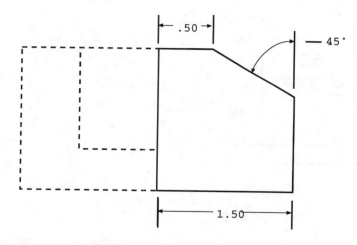

#PROTRUSION
#DONE
#DONE
!Select option for the TO end of the feature.
#UPTO SURFACE
#DONE
!Select or create a SKETCHING PLANE.
#QUERY SEL
!Hidden geometry may be selected.
!Arrow shows direction of feature creation. Pick FLIP or OKAY.
#FLIP
#OKAY
!Select or create a horizontal or vertical reference for sketching.
#TOP
#GEOM TOOLS
#USE EDGE

#QUERY SEL
#DIMENSION
#REGENERATE
!Regeneration completed successfully
!Enter a new value [0.4900]:
.5
!Enter a new value [44.4200]:
45
!Enter a new value [1.3000]:
1.5
#REGENERATE
!Regeneration completed successfully.
#DONE
!PROTRUSION has been created successfully.
#VIEW
#DEFAULT
#DONE
#FEATURE
#CREATE

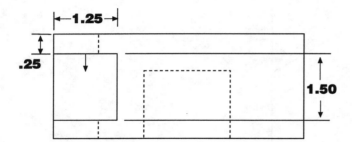

#CUT
#DONE
#DONE
!Select option for the TO end of the feature.
#THRU ALL
#DONE
!Select or create a SKETCHING PLANE.
#MAKE DATUM
!Select an option from the DATUM PLANE menu.
#OFFSET
!Select one of the following: Plane, Coordinate System.
#QUERY SEL
#ENTER VALUE
!Enter offset in the indicated direction, <ESC> to quit [0.15]:
-.25
!Datum Plane is fully constrained. Select "Done," "Quit," or "Restart."
#DONE
!Arrow shows direction of feature creation. Pick FLIP or OKAY.
#OKAY

#ALIGNMENT
!-- ALIGNED --
#DIMENSION
#REGENERATE
!Regeneration completed successfully.
!Enter a new value [0.2700]:
.25
!Enter a new value [1.4800]:
1.5
!Enter a new value [1.1900]:
1.25
#REGENERATE
!Regeneration completed successfully.
#DONE
!Arrow points TOWARD area to be REMOVED. Pick FLIP or OKAY.
#FLIP
#OKAY
!CUT has been created successfully.
#VIEW
#DEFAULT
#CREATE

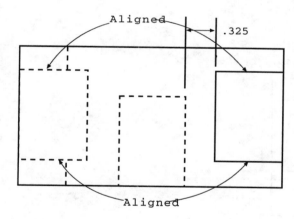

#CUT
#DONE
#DONE
!Select option for the TO end of the feature.
#THRU ALL
#DONE
!Select or create a SKETCHING PLANE.
#MAKE DATUM
!Select an option from the DATUM PLANE menu.
#OFFSET
!Select one of the following: Plane, Coordinate System.
#QUERY SEL

!Hidden geometry may be selected.
#ENTER VALUE
!Enter offset in the indicated direction, <ESC> to quit [0.15]:
-.325
!Datum Plane is fully constrained. Select "Done," "Quit," or "Restart."
#DONE
!Arrow shows direction of feature creation. Pick FLIP or OKAY.
#OKAY
#ALIGNMENT
#DIMENSION
#QUERY SEL
#REGENERATE
!Regeneration completed successfully.
!Enter a new value [0.3100]:
.325
#REGENERATE
!Regeneration completed successfully.
#DONE
!Arrow points TOWARD area to be REMOVED. Pick FLIP or OKAY.
#OKAY
!CUT has been created successfully.
#VIEW
#DEFAULT
#CREATE

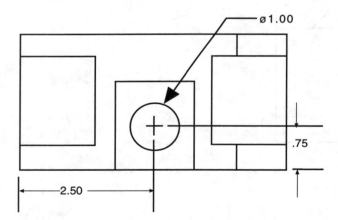

#PROTRUSION
#DONE
#DONE
!Select option for the TO end of the feature.
#DONE
!Select or create a SKETCHING PLANE.
#OKAY
!Select or create a horizontal or vertical reference for sketching.
#DIMENSION
#REGENERATE
!Regeneration completed successfully.

!Enter a new value [0.6000]:
1
!Enter a new value [0.6000]:
.75
!Enter a new value [2.4000]:
2.5
#REGENERATE
!Regeneration completed successfully.
#DONE
!Enter depth [1.48]:

!PROTRUSION has been created successfully.
#VIEW
#DEFAULT
#REROUTE
!Select feature to reroute.
!Select an alternate sketching plane.
!Showing surface created by feature 1 (FIRST FEATURE).Confirm
selection.
#SAME REF
!Select an alternate dimensioning reference.
#SAME REF
!Select an alternate dimensioning reference.
#SAME REF
!Feature rerouted successfully.
#DONE
#MODIFY
!Select FEATURE or DIMENSION.
!Enter depth [1.48]:
1.25
#REGENERATE
#FEATURE
#CREATE

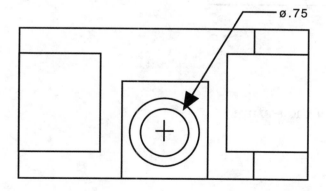

#PROTRUSION
#DONE
#DONE
!Select option for the TO end of the feature.
#DONE
!Select or create a SKETCHING PLANE.
#QUERY SEL
#FLIP
#OKAY
#CIRCLE
#CONCENTRIC
#DIMENSION
#REGENERATE
!Enter a new value [0.7400]:
.75
#REGENERATE
!Regeneration completed successfully.
#DONE
!Enter depth [1.48]:
1.5
!PROTRUSION has been created successfully.
#VIEW
#DEFAULT
#CREATE

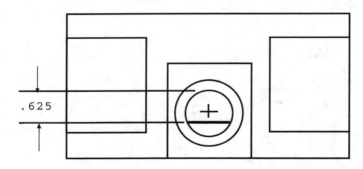

#CUT
#DONE
#DONE
!Select option for the TO end of the feature.
#UPTO SURFACE
#DONE
#OKAY
#ALIGNMENT
#DIMENSION
#TANGENT
#REGENERATE
!Regeneration completed successfully.

!Enter a new value [0.6200]:
.625
#REGENERATE
!Regeneration completed successfully.
#DONE
!Arrow points TOWARD area to be REMOVED. Pick FLIP or OKAY.
#FLIP
#OKAY
!CUT has been created successfully.
#VIEW
#DEFAULT
#CREATE
#DONE/RETURN
#DONE
#X-SECTION
!Select a cross section to retrieve by picking names on the menu.
#CREATE
#DONE
!Enter NAME for cross-section [QUIT]:
a
!Select planar surface or datum plane.
#DONE/RETURN

Answer Key to Exercise 4-2

Exercise 4-2:	Create the base feature shown in Figure 4-8. Take care to follow the dimensioning scheme shown. This is a lead-in to the next exercise. When finished, proceed to Exercise 4-3.

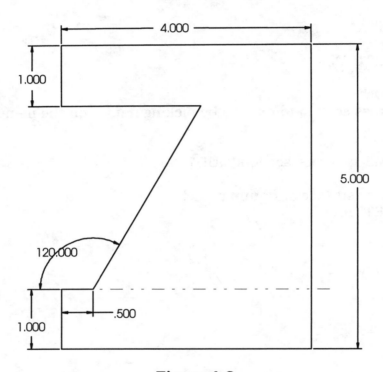

Figure 4-8.

The following are the menu picks:

> *Part/Create/"block"*
> *Feature/Create/Protrusion/Done*

Sketch as shown in Figure 4-8 and modify the dimensions to the values shown.

> *Regenerate/Done/1.0 <CR>*

Answer Key to Exercise 4-3

Exercise 4-3:	After the part is extruded, set the 120-degree angle to 90 degrees. Create the protrusion shown in Figure 4-9, being careful to choose the sketching plane and vertical reference plane as indicated.

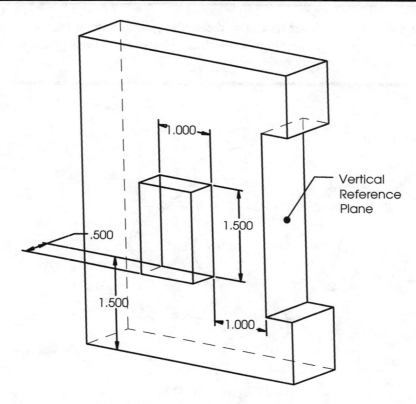

Figure 4-9.

The menu picks, starting at the PART menu, are as follows:

> *Modify.* Pick the base feature and change the 120-degree angle to 90 degrees.
> *Regenerate/Feature/Create/Cut/Done/Done/Done*

Select the front face of the part as the sketching plane and the vertical reference as shown in Figure 4-9. Sketch as shown in Figure 4-9.

> *Regenerate/Done/Flip/Okay/0.50 <CR>*
> *View/Default/Done-Return*

Now make the change to the angle.

> *Modify.* Pick the base feature and change the 90-degree angle to
> 120 degrees.
> *Regenerate*

Answer Key to Exercise 4-4

Exercise 4-4:	Your task is to create the gear shown in Figure 4-10A by patterning the tooth angularly around the perimeter of the gear body.

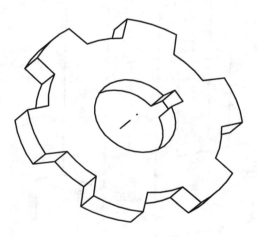

Figure 4-10A.

Step 1:	Create the gear body and dimension it as shown in Figure 4-10B.

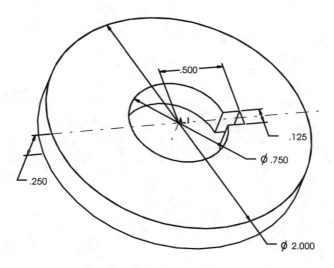

Figure 4-10B.

Step 2:	Prepare to create the *Thru All* cut that removes material for the first tooth by choosing the surface shown in Figure 4-10C as the sketching plane. When asked to choose a Vert & Horiz Reference choose *Make Datum* and constrain the datum "through" the axis and "angled" to the edge of the key-way (see Figure 4-10C). Enter 45 degrees as the angle for the plane.

Questions:

After entering the 45-degree angle, Pro/ENGINEER dropped you into the Sketcher. Why?

The sketching plane had been defined as the gear face and the *Make Datum* is used to orient the gear into sketches.

Why is the datum on the fly facing the way it is?
Orientation of the datum plane is based on the order of selecting references as well as specifying Top, Bottom, etc.

Is there another datum shown? If so why?
Yes, it is the vertical orientation plane that is normal to the datum on the fly.

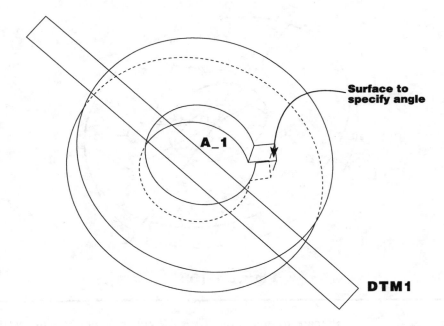

Surface to specify angle

A_1

DTM1

Figure 4-10C.

Step 3:		Construct the cut for the first tooth as shown below.
	1.	Under the SEC TOOLS menu, select *Retrieve, Sec, ?*, and choose an item.
	2.	For the rotating angle, use the default (0.0). When asked to select origin point for scaling, choose one end of sketch. When asked to select point to move, select the other end.
	3.	For scaling factor, use the default (1.0).
	5.	When asked to place section on part, do so by dragging the section into place. (Use *Front* key to change windows.)
	6.	Now align both ends and regenerate. When the sketch regenerates, choose *Done* and extrude the cut.
	7.	Your construction should now look like the following figure.

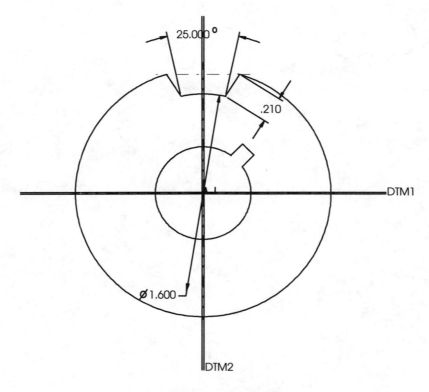

Figure 4-10D.

Step 4:	Choose *Modify,* pick the feature, and notice that the 45-degree angle is now associated with the cut. Pattern the cut around the gear six times on 60-degree centers.

The following are the steps to follow in order to pattern the cut.

From the PART menu select:

   Feature/Pattern/"pick the cut feature"*/Done*

Select the 45-degree dimension as the first direction.
Enter an increment of 60. Select *Done.*
Enter 6 for the number of instances. Select *Done.*

The pattern has been created.

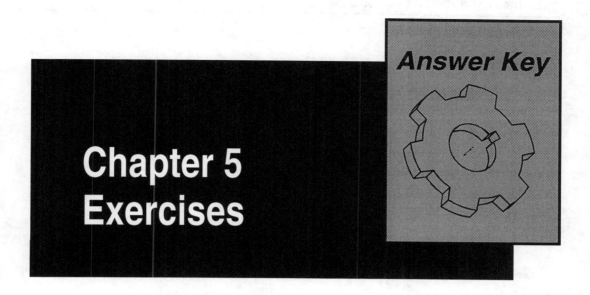

Answer Key

Chapter 5
Exercises

Answer Key for Exercise 5-1

Exercise 5-1: Creating a generic part trail file:

1. Exit and restart Pro/ENGINEER.

2. Create a part made up of the default datum planes.

3. Define and save the following views: Front, Back, Right, Left, Top, and Bottom.

Note: The yellow side of the datum goes to the direction selected in the menu.

4. Select DBMS, Rename, <CR>, <CR>, then exit Pro/ENGINEER.

5. Pull up a shell window and type *ls*. Note the full name of the latest *trail.txt* file and copy it to a file with a different name by typing *cp trail.txt.<number> startpart.txt*.

6. Edit this file with the 'Text edit' editor. Go to the end of the trail file and delete all lines after: !Enter TO: , and then save the file.

7. Log into Pro/ENGINEER and select *Misc*, *Trail*, and enter the name of your trail file at the prompt. After the trail file has executed completely, you can type in the new part name when prompted, and you're ready to build your model. You can also run a trail file when starting Pro/ENGINEER by adding the trail file name to the Pro/ENGINEER start-up command.

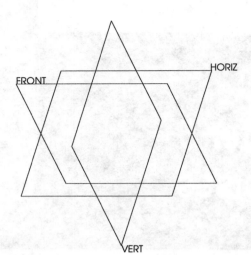

Figure 5-2.

To create a generic start part follow these steps:

> *Part/Create/Begin<CR>*
> *Feature/Create/Datum/Plane/Default*
> *Done/Set Up/Name*

Change the names of the datum planes as shown in Figure 5-2.

> *Done/View/Front.* Select the "FRONT"datum.
> Top. Select the "HORIZ" datum.
> *Names/Save/"Front" <CR>/Orientation/Default*
> Back. Select the "FRONT" datum.
> Top. Select the "HORIZ" datum.
> *Names/Save/"back" <CR>/Orientation/Default*
> Right. Select the "VERT" datum.
> Top. Select the "HORIZ" datum.
> *Names/Save/"right" <CR>/Orientation/Default*
> Left. Select the "VERT" datum.
> Top. Select the "HORIZ" datum.
> *Names/Save/"left" <CR>/Orientation/Default*
> Front. Select the "HORIZ" datum.
> Bottom. Select the "FRONT" datum.
> *Names/Save/"top" <CR>/Orientation/Default*
> Back. Select the "HORIZ" datum.
> Bottom. Select the "FRONT" datum.
> *Names/Save/"bottom" <CR>/Orientation/Default*
> *DBMS/Rename/<CR>/<CR>*
> *Exit*

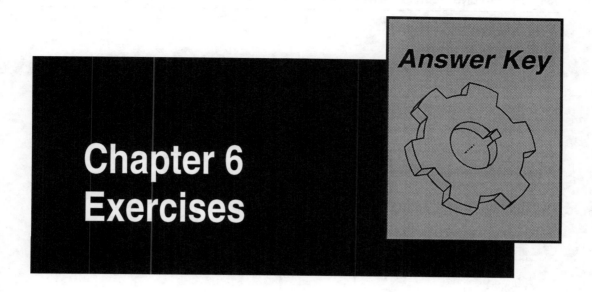

Chapter 6 Exercises

Answer Key for Exercise 6-1

Exercise 6-1:   Create the part shown in Figure 6-6.

1. Begin with the default datum planes.
2. Create the block using the dimensions shown. Position the block so that two datum planes pass through the position of the first hole (lower left-hand corner).
3. Now create the hole. Select *Feature, Create, Solid Hole, Dim Pattern, Linear, Done*. Select *Thru All* and enter the diameter.
4. When prompted to select the placement plane, select a point above and to the right of the two datum planes. Select the two datums for dimensioning and accept the default dimensions.
5. Select *General* and *Done*. From the PAT DIM INER menu, select *Table*, then select the two dimensions from step 4. *Done*. From the PATT TABLE menu, select *Add* and enter the name for the pattern table.
6. When the Pro/TABLE displays, type in the following values.
   Table Name XXXXXXX

idx	d1	d2	(note your particular
1	6.25	0	dimension's orientation)
2	12.50	0	
3	0	3.25	
4	12.50	3.25	
5	0	6.5	
6	6.25	6.5	
7	12.5	6.5	

7. Exit the Pro/TABLE editor and select *Done* and *Done*.
8. Now modify your first hole's dimensions back to zero. Store the part for later use.

   Do you get the dimensioning scheme shown in Figure 6-6?

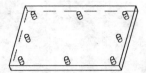

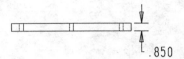

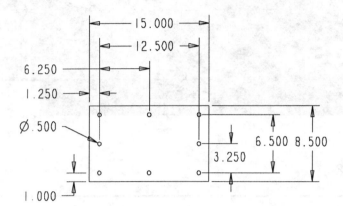

Figure 6-6.

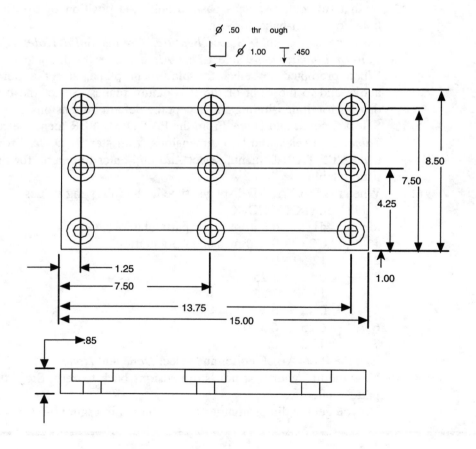

1.   *Part/Create/"block"*
      Feature/Create/Protrusion/Done
      Sketch the block profile.
      Dimension/Regenerate/Done

2.   *Feature/Create/Hole/Done*
      Thru All/Done
      Enter the diameter 0.50. Place on the front surface.
      Dimension from the bottom edge or surface and from the left edge
      or surface.

3.   *Feature/Create/User Defined/Search/Retr*
      Retrieve a counterbore feature from disk and place it on the part.

4.   *Feature/Group/Ungroup/Done*

5.   *Feature/Group/Local Group*
      Enter a name for the local group.
      Select the hole and the counterbore to group.
      Done

6.   *Pattern* from the GROUP menu.
      Value. Select the 1.00 dimension, increment by 3.25.
      Done
      Enter the total number of instances as 3.
      Value. Select the 1.25 dimension, increment by 6.25.
      Done
      Enter the total number of instances as 3.

7.   *Feature/Group/Unpattern.* Select one of the holes.
      Done

8.   *Feature/Delete.* Select the center hole.
      Done

Now select *Modify* to show all dimensions.

Answer Key for Exercise 6-2

Exercise 6-2:

1. Create the generic screw following the dimensioning scheme shown in Figures 6-7 and 6-8.

2. From the PART menu, select *Family Tab, Add Item, Dimension,* select the dimensions in order, then select *Done/Return.*

3. To include descriptive names of the symbolic dimensions in the table, choose *Modify, Dim Cosmetics, Symbol* and pick each dimension. Type in the symbolic name for that dimension. See the table in Figure 6-9. Continue naming all dimensions as shown in the table. Select *Done, Done.*

4. Go to the FAMILY TABLE menu, select *Edit,* and add the information shown in Figure 6-9 to your Generic Screw Table. Be careful when typing in data.

5. Exit the Pro/TABLE Editor.

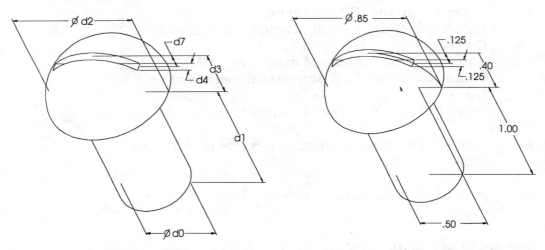

Figure 6-7.   **Figure 6-8.**

	d0	d1	d2	d3	d4	d7
	thd_dia	thd_len	head_dia	head_ht	slot_wid	slot_dep
Generic	.500	1.000	.850	.400	.125	.125
Screw_1	.480	.987	.800	.380	.120	.120
Screw_2	*	*	*	*	*	*
Screw_3	.250	.750	.420	.200	.120	.120

Figure 6-9.

1.　　Create the generic screw:

　　　Part/Create/"screw" <CR>
　　　Feature/Create/Protrusion/Done

Sketch a circle and dimension the diameter with the value shown in Figure 6-8.

　　　Regenerate/Done/1.0 <CR>
　　　Create/Protrusion/Revolve/Done/360/Done
　　　Make Datum/Through/pick axis A_1/Done
　　　Okay/Top/Pick the cylinder top

Sketch as shown:

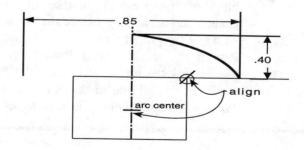

Regenerate/Done/Okay
Feature/Create/Done/Done/Both Sides/Blind/Done
Make Datum/Through/Pick axis A_1/Done
Okay/Bottom/Pick the cylinder bottom

Sketch a line as shown:

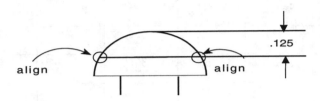

Regenerate/Done/Okay (arrow points up)
.125 <CR>/*Done*

2.　Create the family table and select the dimensions in the order shown in Figure 6-9.

5.　Once you have an instance displayed in a member window, use *Quit Window* to remove the instance display, then *Change Window*, picking the main window to activate it.

Exercise 6-2 (continued):

6.　To view an instance, choose *Instance* and select an instance listed in the Pop-up menu. *Quit Window* to exit the small window. *Change Window* and select on the large graphic window to make it active. Store the part and *Quit Window*.

7.　In the assembly mode create an assembly and retrieve the block of Figure 6-6. Then assemble instance Screw_3 of the generic screw part into one of the holes. Too small? Replace it with one the proper size.

8.　Pattern the screw. Select *Pattern* from the COMPONENT Menu, pick the screw, and select *Ref Pattern*, and *Done*. The screw will be placed in the hole pattern. If you selected *Ref Pattern* in step 7, there is no need to do so for step 8.

Note: Because you unpatterned the holes in order to delete the center hole, there is no pattern for you to use with the screw. Therefore, you have to insert each screw individually.

6.   Create the assembly:

Assembly/Create/"plate" <CR>
Component/Assemble/Enter the name of the part from 6-5 <CR>
Assemble/"screw" <CR> Answer yes to the instance prompt
* <CR>/*SCREW_3/Done*
Insert/Pick the screw cylinder, then pick the inside cylindrical surface of one of the holes in the plate.
Mate/Pick the face of the plate and the underside of the screw head/*Done*

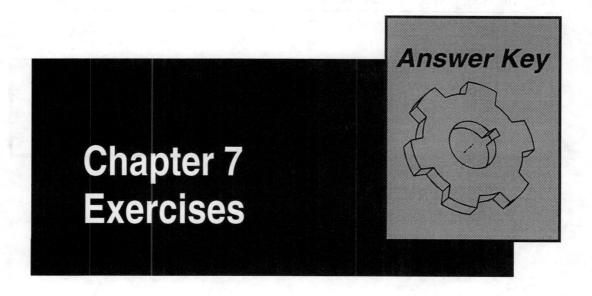

Chapter 7
Exercises

Answer Key for Exercise 7-1

Exercise 7-1	The part shown below was designed and modeled using poor modeling practices. Management wants it corrected because it is causing problems in manufacturing. You are given the job of fixing what someone else modeled without knowing how he or she did it. Follow the tasks outlined. **Note:** There may be a geometry check in this exercise. This is only a warning, but if one is found, it must be corrected.

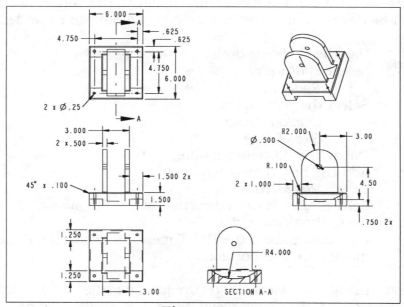

Figure 7-4.

> **Task 1:**   Modify the part to remedy the driven dimensions (the 4.50-inch dimension in the right drawing view and the 1.500-inch dimension in the front view) and make them feature or model dimensions.

---

1.   Retrieve the part model *grind.prt*.

2.   To identify the driven dimensions:

    a.   Retrieve the drawing *grind_base.dwg*. (The driven dimensions should highlight in red.)

    b.   Note which dimensions are highlighted; these are the driven dimensions that you will convert to model dimensions. If no dimensions highlight, select *Regenerate* and *Draft*.

    c.   Quit the drawing window.

3.   Retrieve the part again using *Search/Retr* and *In Session*. Suppress unneeded features to clarify the part and speed regeneration time. Be sure that you suppress the center cut, the hole through the protrusion, the rounds, and the chamfers.

4.   Now consider the 4.50-inch driven dimension in the right drawing view. Redefine the protrusions cross-section to make the 4.50-inch driven dimension a feature (model) dimension. (Note that the present 3.00 vertical model dimension to the center of the radius will be deleted and a new dimension will have to be added.)

    Use the *Redefine* option.
    Select *Feature*
        Redefine
    Select the protrusion
        Section/Done
        Sketch
    Delete the 3.00 dimension.
    Add the new dimension.
        Regenerate
    Select *Done*, set the direction of the flip arrow.
    Go to the default view
    Use *Modify* from the PART menu to check the dimension to make sure it is correct.

5.   Next, using the *elevator* technique, fix the 1.500-inch driven dimension in the front view. Start by creating a datum plane

coincident with the near face of the base, offset from datum left 6.00 inches.

> Select *Create* from the FEAT menu.
> > *Datum*
> > *Plane*
> > *Done*
> > *Offset*

Select one of the following: plane, coordinate system.
> Select datum left

Select location on model for offset value or choose Enter value from menu.
> Enter value

Enter offset in the indicated direction,<Esc> to quit [0.27]:
Check the direction of the green arrow
> Enter *6.00/Done*

Datum DTM8 is created coincident to the near face of the base feature.

6.   Reorder DTM8 to be an earlier feature.

7.   Reroute the 1.50 dimension to datum DTM8.

> Select *Reroute*
> Select the protrusion.

"Do you want to roll back the part? (N)" *Y* <CR>
When the near face highlights in purple, select *Alternate* and select Datum DTM8.
For all other surfaces highlighted, select *Same Ref.*

8.   Change the datum DTM8 offset dimension to 0.0. This makes DTM8 coincident with datum left.

> Select *Modify* from the PART menu.
> Select datum DTM8.
> Change the 6.00 dimension to 0.0.
> *Regenerate*

Note: The protrusion is now tied to datum DTM8, but it displays off the base feature.

9.   Reverse the direction of the 1.50 dimension (-1.50).

> Select *Modify*
> Select the protrusion.
> Change the 1.500 dimension to -1.500.
> *Regenerate*

The protrusion in now on the base feature, but it is not yet in the correct position.

10.   Position the protrusion correctly by redefining the direction the protrusion was created.

> Select *Redefine* from FEAT menu.
> Select the protrusion
> Select *Direction/Done*
> > *Flip/Okay*

The protrusion is now in the correct position.

11.   Reroute the protrusion back to datum left.

> Select *Reroute* from the FEAT menu.
> Select the protrusion.
> "Do you want to roll back the part? (n)" *Y <CR>*
> When DTM8 high lights, select *Alternate*.
> > *Query Sel*
> Select on *DTM8/Next* until DTM LEFT highlights.
> > *Accept*
> Select *Same Ref* for all other highlighted surfaces.

12.   Delete datum DTM8. The driven dimensions have now been corrected.

13.   Resume all suppressed features.

Task 2:	There are nine drilled holes in the base feature. Five of these are plugged. Correct this situation so only the four corner holes remain.

---

1.   Suppress all features except the base feature, the holes, and the cut in the bottom of the base.

2.   Turn hidden lines on and datums off.

3.   Modify the hole pattern so only the corner holes remain.

> Select *Modify* from the PART menu.
> Select the lower right corner hole.
> Change the 2.375 dimensions to 4.750 (2.375*2).
> Change the 3 HOLES count to 2 HOLES.
> > *Regenerate*
> "Cannot delete pattern member with children."

"Red feature couldn't regenerate. Select recovery option from menu."

   Select *Restore All* from REGEN FAIL menu.

Note: The plugs must be deleted first.

4.   Determine the feature number of the plugs and delete them by feature number.

   Select *Info* from the PART menu.
       Regen Info
       Beginning
   Continue watching each feature as it is displayed, until all features are displayed. Note the feature number of the plugs.
   The feature numbers are displayed on the message line.
   Select *Info Feat* for additional feature information.
   Note the feature number of the plugs (protrusions).
   Quit

5.   Delete the plugs.

   Select *Feature* from the PART menu.
       Delete
       Sel By Menu
       Number
   "Enter feature regeneration number(1-16)[QUIT]:"
   Enter the feature number for the plugs.
   Select *Done* from the SELECT FEAT menu.

Note: All nine holes are now displayed.

6.   Now modify the hole pattern to 2 x 2 per instructions in step 3.

What happened? "Cannot delete pattern member with children."
"Red feature could not regenerate. Select recovery option from window."

   Select *Restore All* from the REGEN FAIL menu.

7.   Determine what feature is the child and reroute it.

   Select *Feature* from the PART menu.
       Suppress
   Select the upper middle hole.

Note: The bottom cuts are children of the holes.
       "Select option for child in blue"
       Select *Reroute*
       "Do you want to roll back the part?[N]:" *Y<CR>*

Redefine the cuts in the base.
Select *Sketch.*
Unalign the cyan centerline from the middle row of holes and dimension it to the front face of the base.
Regenerate
Done
Now modify the holes.

Task 3:	Modify the 3-inch dimension between the wheel support to 4 inches, while keeping the protrusion centered on the base feature.

1.   Suppress all unneeded features.

2.   Modify the required dimensions.

Select *Modify* from the PART menu.
Select the protrusion and cut features.
Make the necessary dimension changes.

Task 4:	Modify the tolerance of the base feature, 6" x 6" x 1.5", to two decimal places. Experiment with the other tolerance options.

1.   Suppress all features except the base feature.

2.   Using Environment set *TOL ON.*

3.   Display the base feature dimensions.

4.   Modify the tolerances to two decimal places.

Select *Modify* from the PART menu.
    Dim Cosmetics
    Format
    Limits
Select on the dimensions.
    Done Sel
    Num Digits
"Number of decimal places for dimensions (2)" <CR>
Select dimensions; they will change as selected.

5.   Resume all features.

Task 5:	Make the 4-hole location dimensions basic dimensions and add a geometric tolerance to the 0.25 diameter hole location. Make this a true position of 0.010 inch relative to datums TOP, FRONT, and LEFT.

1.   Suppress all features but the base feature and the holes.

2.   Turn datums on.

3.   Make the four-hole location dimensions basic dimensions.

> Select *Modify* from the PART menu.
> Display hole dimensions.
> Select *Setup*
> > *Geom Tol*
> > *Basic Dim*
> "Select dimensions which are basic for tolerancing."
> Select the four locating dimensions.
> Select *Done Sel*
> > *Done*
> > *Done/Return*

4.   Convert datum planes to geometric tolerance datums.

> Turn datums on.
>
> Select *Set Up* from the PART menu.
> > *Geom Tol*
> > *Set Datum*
> "Select reference datums."
> Select a datum plane: FRONT, TOP, LEFT.
> Select *Default*
> Repeat until all datums have been set.
> Select *Done/Return*
> > *Done*

5.   Add a geometric tolerance to the 0.25-inch diameter hole dimension.

> Select *Setup* from the PART menu.
> > *Geom Tol*
> > *Specify Tol*
> > *Location*
> > *Position*
> > *Feature*

"Select feature."
Select the hole.
>   *MMC*

"Enter tolerance value [0.001]:" .01<CR>
"Select whether the tolerance value refers to a diameter zone or a width."
Select *Diameter*
"Pick datums to which the selected entity is referenced."
(Use *Compound* to reference more than one datum.)
Pick the three datums and the material condition, MMC
>   *Done Sel*
>   *Done*

"This feature base has some external features which are non-basic. Convert?(Y/N):" *N* <CR>
Select *Dim*
Select the hole diameter dimension.
>   *Done/Return*

Index

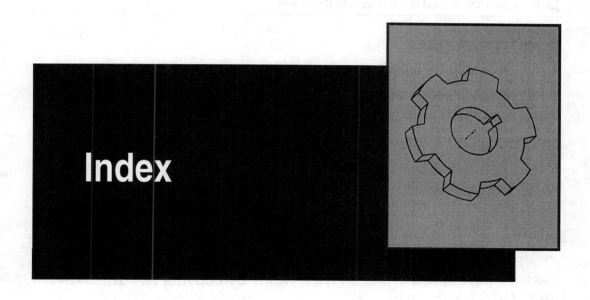

More OnWord Press Titles

Pro/ENGINEER Books
INSIDE Pro/ENGINEER
Book $49.95 Includes Disk

The Pro/ENGINEER Quick Reference
Book $24.95

The Pro/ENGINEER Exercise Book
Book $39.95 Includes Disk

Interleaf Books
INSIDE Interleaf
Book $49.95 Includes Disk

The Interleaf Quick Reference
Book $24.95

The Interleaf Exercise Book
Book $39.95 Includes Disk

Adventurer's Guide to Interleaf Lisp
Book $49.95 Includes Disk

Interleaf Tips and Tricks
Book $49.95 Includes Disk

MicroStation Books
INSIDE MicroStation 5X. Third Edition
Book $34.95 Includes Disk

MicroStation Bible
Book $49.95 Optional Disk $49.95

MicroStation Reference Guide 5.X
Book $24.95 Includes Disk

MicroStation for AutoCAD Users 5.X
Book $34.95 Includes Disk

MicroStation 5.X Delta Book
Book $19.95

The MicroStation Productivity Book
Book $39.95 Optional Disk $49.95

Adventures in MDL Programming
Book $49.95

Programming With MDL
Book $49.95 Optional Disk $49.95

Programming With User Commands
Book $65.00 Optional Disk $40.00

101 MDL Commands
Book $49.95 Optional Executable Disk $101.00
Optional Source Disks (6) $259.95

101 User Commands
Book $49.95 Optional Disk $101.00

Bill Steinbock's Pocket MDL Programmers Guide
Book $24.95

MicroStation for AutoCAD Users Tablet Menu
Tablet Menu $99.95

Managing and Networking MicroStation
Book $29.95 Optional Disk $29.95

The MicroStation Database Book
Book $29.95 Optional Disk $29.95

The MicroStation Rendering Book
Book $34.95 Includes Disk

INSIDE I/RAS B
Book $24.95 Includes Disk

Build Cell
Software $69.95

The CLIX Workstation User's Guide
Book $34.95 Includes Disk

SunSoft Solaris Series
The SunSoft Solaris 2.* User's Guide
Book $29.95 Includes Disk

SunSoft Solaris 2.* for Managers and Administrators
Book $34.95 Optional Disk $29.95

The SunSoft Solaris 2.* Quick Reference
Book $18.95

Five Steps to SunSoft Solaris 2.*
Book $24.95 Includes Disk

One Minute SunSoft Solaris Manager
Book $14.95

SunSoft Solaris for Windows Users
Book $24.95

The Hewlett Packard HP-UX Series
The HP-UX User's Guide
Book $29.95 Includes Disk

HP-UX for Managers and Administrators
Book $34.95 Optional Disk $29.95

The HP-UX Quick Reference
Book $18.95

Five Steps to HP-UX
Book $24.95 Includes Disk

One Minute HP-UX Manager
Book $14.95

HP-UX for Windows Users
Book $24.95

CAD Management
One Minute CAD Manager
Book $14.95

Manager's Guide to Computer-Aided Engineering
Book $49.95

Other CAD
CAD and the Practice of Architecture: ASG Solutions
Book $39.95 Includes Disk

Fallingwater in 3D Studio: A Case Study and Tutorial
Book $34.95 Includes Disk

INSIDE CADVANCE
Book $34.95 Includes Disk

Using Drafix Windows CAD
Book $34.95 Includes Disk

Management
How to Stay Stressed
Book $9.95

Geographic Information Systems
The GIS Book. Third Edition
Book $39.95

DTP/CAD Clip Art
1001 DTP/CAD Symbols Clip Art Library: Architectural
Book $29.95
MicroStation
DGN Disk $175.00 Book/Disk $195.00
AutoCAD
DWG Disk $175.00 Book/Disk $195.00
CAD/DTP
DXF Disk $195.00 Book/Disk $225.00

Networking/LANtastic
Fantastic LANtastic
Book $29.95 Includes Disk

The LANtastic Quick Reference
Book $14.95

One Minute Network Manager
Book $14.95

OnWord Press Distribution

End Users/User Groups/Corporate Sales
OnWord Press books are available worldwide to end users, user groups, and corporate accounts from your local bookseller or computer/software dealer, or from HMP Direct: call 1-800-526-BOOK or 505-473-5454; fax 505-471-4424; e-mail to ORDERS@BOOKSTORE.HMP.COM; or write to High Mountain Press Direct, 2530 Camino Entrada, Santa Fe, NM 87505-4835.

Wholesale, Including Overseas Distribution
We have international distributors. Contact us for your local source by calling 1-800-ONWORD or 505-473-5454; fax to 505-471-4424; e-mail to ORDERS @BOOKSTORE.HMP.COM; or write to High Mountain Press/IPG, 2530 Camino Entrada, Santa Fe, NM 87505-4835, USA.

Comments, Corrections, and Bug Fixes

Your comments can help us make better books. If you find an error in our products, or have any other comments, positive or negative, we'd like to know! Please contact our e-mail address: READERS@HMP.COM, or write to us at the address below.

OnWord Press
2530 Camino Entrada
Santa Fe, NM 87505-4835 USA